Prentice Hall Advanced Reference Series 

Physical and Life Sciences

BINKLEY *The Pineal: Endocrine and Nonendocrine Function*
CAROZZI *Carbonate Rock Depositional Models: A Microfacies Approach*
EISEN *Mathematical Methods and Models in the Biological Sciences: Linear and One-Dimensional Theory*
EISEN *Mathematical Methods and Models in the Biological Sciences: Nonlinear and Multidimensional Theory*
FRASER *Stratigraphic Evolution of Clastic Depositional Sequences*
JEGER, ED. *Spatial Components of Plant Disease Epidemics*
MCLENNAN *Introduction to Nonequilibrium Statistical Mechanics*
PLISCHKE AND BERGERSEN *Equilibrium Statistical Physics*
SORBJAN *Structure of the Atmospheric Boundary Layer*
VALENZUELA AND MYERS *Adsorption Equilibrium Data Handbook*
WARREN *Evaporite Sedimentology: Importance in Hydrocarbon Accumulation*

Spatial Components of Plant Disease Epidemics

Michael J. Jeger, Editor

Overseas Development
Natural Resources Institute
London, England

Prentice Hall, Englewood Cliffs, New Jersey 07632

Library of Congress Cataloging-in-Publication Data

Spatial components of plant disease epidemics / MICHAEL J. JEGER, editor.

p. cm.—(Prentice Hall advanced reference series)
Includes bibliographies and index.
ISBN 0-13-824491-X
1. Plant diseases—Epidemiology. 2. Spatial analysis (Statistics)
3. Phytogeography. I. Jeger, Michael J.
QK731.S75 1989
581.2′4—dc19 88-13081
CIP

Editorial/production supervision
and interior design: *Jean Lapidus*
Cover design: *Karen Stephens*
Manufacturing buyer: *Mary Ann Gloriande*

Prentice Hall Advanced Reference Series

Printed in the United States of America
10 9 8 7 6 5 4 3 2 1

ISBN 0-13-824491-X

PRENTICE-HALL INTERNATIONAL (UK) LIMITED, *London*
PRENTICE-HALL OF AUSTRALIA PTY. LIMITED, *Sydney*
PRENTICE-HALL CANADA INC., *Toronto*
PRENTICE-HALL HISPANOAMERICANA, S.A., *Mexico*
PRENTICE-HALL OF INDIA PRIVATE LIMITED, *New Delhi*
PRENTICE-HALL OF JAPAN, INC., *Tokyo*
SIMON & SCHUSTER ASIA PTE. LTD., *Singapore*
EDITORA PRENTICE-HALL DO BRASIL, LTDA., *Rio de Janeiro*

Contents

Contributors

P. B. ADAMS, Biocontrol of Plant Diseases Laboratory, Plant Sciences Institute, U.S. Department of Agriculture, Beltsville, MD 20705, USA.

H. M. ALEXANDER, Department of Biology, University of Louisville, Louisville, KY 40292, USA. *Current address:* Departments of Botany and Systematics & Ecology, University of Kansas, Lawrence, KS 66045, USA.

D. N. APPEL, Department of Plant Pathology and Microbiology, Texas A & M University, College Station, TX 77843, USA.

P. H. BERGER, Department of Plant Pathology, University of Kentucky, Lexington, KY 40546, USA. *Current address:* Department of Plant, Soil, and Entomological Sciences, University of Idaho, Moscow, ID 83843, USA.

R. S. FERRISS, Department of Plant Pathology, University of Kentucky, Lexington, KY 40546, USA.

R. E. GAUNT, Department of Agricultural Microbiology, Lincoln University College of Agriculture, Canterbury, New Zealand.

J. V. GROTH, Department of Plant Pathology, University of Minnesota, St. Paul, MN 55108, USA.

M. E. IRWIN, Illinois Natural History Survey and University of Illinois, 607 E. Peabody, Champaign, IL 61820, USA.

M. J. JEGER, Overseas Development Natural Resources Institute, 56–62 Gray's Inn Road, London WC1X 8LU, UK.

G. E. KAMPMEIER, Illinois Natural History Survey and University of Illinois, 607 E. Peabody, Champaign, IL 61820, USA.

L. V. MADDEN, Department of Plant Pathology, The Ohio State University, Ohio Agricultural Research and Development Center, Wooster, OH 44691, USA.

D. MARSHALL, Texas Agricultural Experiment Station, Texas A & M University Research and Extension Center, Dallas, TX 75252, USA.

K. P. MINOGUE, Department of Botany and Plant Pathology, Purdue University, West Lafayette, IN 47907, USA.

M. J. ROBERTSON, Department of Agriculture, University of Queensland, St. Lucia, Queensland 4067, Australia.

Preface

In the last 25 years epidemiology has matured as a discipline within the science of plant pathology. Most epidemiological studies, however, whether descriptive or quantitative, have been biased toward temporal aspects of plant disease epidemics: how epidemics develop within and between growing seasons; the time taken for pathogen populations to adapt in response to the release and deployment of resistant cultivars or new fungicides; and the development of tactics and strategies of disease management based on population dynamic models. In many ways this bias mirrors early developments in population ecology in which the temporal attributes of populations received more attention than did spatial attributes. As it did in ecology, the situation is changing in plant disease epidemiology. The spatial constraints on epidemics, whether on a global or a local scale, are now recognized as critical elements in attempts to understand and manage plant disease epidemics to better effect.

Nonetheless, there are few sources in the plant pathology literature in which the importance of spatial considerations has been recognized and coherently discussed. This book, it is anticipated, will serve as such a source, by bringing together contributions from many different areas of plant disease epidemiology, plant ecology, population genetics of plant pathogens, arthropod vector–virus relationships, mathematical modeling, and mainstream plant pathology—united by a common concern with the spatial attributes of epidemics. Several contributions are made by authors who made presentations at a colloquium on spatial components of epidemics held at the 1985 Annual Meeting of the American Phytopathological Society in Reno, Nevada; others have since been solicited to broaden the range of topics covered in this book. The integration of different areas of plant disease epidemiology is a major distinctive feature of this book. Other recent books have been organized in terms of discipline, pathogen group, host, or disease type, but few have attempted integration around a pervasive theme such as that described here.

The book should be suitable for use as a basic text for advanced courses in plant disease epidemiology and management, in combination with other texts, and as a source book for researchers and practitioners in plant pathology. Because of the diverse topics included, the book should also be of interest to plant ecologists, entomologists, applied meteorologists and geographers, and, in general, to all involved with the diverse uses of crop plants in agriculture.

1 The Spatial Component of Plant Disease Epidemics

M. J. Jeger

INTRODUCTION

Plant disease epidemiology is concerned with the determinants and consequences of disease in plant populations. As such, the dynamics of plant disease epidemics in time have often been the major aspect of interest to epidemiologists. Epidemics within a growing season are often spectacular, especially in an annual crop when the host population is similarly increasing in a marked fashion. The temporal increase of plant disease is often the most obvious manifestation of the epidemic. Spatial aspects of plant disease epidemics, by contrast, unless they correspond to the ordered progression of disease over a geographical range, such as the cereal rust epidemics in central United States and Canada, are seldom as obvious unless examined with more discernment.

It is, thus, not surprising that the overwhelming emphasis of the plant-disease-epidemic literature has been on temporal aspects, as perusal of several important sources and textbooks will testify (Vanderplank, 1963; Kranz, 1974; Horsfall and Cowling, 1978; Scott and Bainbridge, 1978; Zadoks and Schein, 1979; Fry, 1982; Plumb and Thresh, 1983; Gilligan, 1985; Leonard and Fry, 1986; Wolfe and Caton, 1987). Although the movement and dispersal of plant

Note: Overseas Development Natural Resources Institute, 55–62 Gray's Inn Road, London WC1X 8LU, UK.

pathogens has received some attention (Gregory, 1968), most recently together with pests and vectors (MacKenzie et al., 1985), it has often been divorced from disease. Part of the reason for this emphasis in plant disease epidemiology may be due to the relative lack of concern for the individual plant, unlike the situation in human and animal epidemiology. It is quite striking, for example, that questions concerning the spread of measles from one side of a major city, Glasgow, to the other side (Bailey, 1975) should seem quite natural in the domain of human epidemiology, yet not possess the same piquancy in the context of a plant disease. Similarly, the fact that plants are relatively immobile, to a large extent, in agricultural crops, uniform and synchronous in development, and usually managed as a single unit can lead to a disregard for the spatial determinants and consequences of disease. It is not by chance, for example, that spatial perceptions and insights are most commonly found in forest tree pathology, especially in the urban and natural environments where individual trees have real value, as will be discussed later.

Several recent developments demonstrate quite clearly the limitations and inadequacies of a purely temporal view of plant disease epidemics, some of these are introduced and discussed in detail in the chapters of this book. A specifically spatial perspective offers new insights and possibilities and areas of research, modeling, and application that can reinvigorate epidemiology and combine both intellectual and practical challenges to the discipline. It is the purpose of this introductory chapter to provide an overview of some of these recent developments and to outline the more substantial chapters that follow. It is also our intention to introduce other important spatial aspects of epidemics that are not so well represented in this book, largely because they have been adequately covered elsewhere in recent years.

DISPERSAL OF PLANT PATHOGENS AND DISEASE VECTORS

Airborne Dispersal

The obvious area that has been intentionally omitted is that of the dispersal of airborne spores of plant pathogens. Since the pioneering work of Gregory (1968), there have been numerous studies on the means by which airborne fungal spores are liberated from crops, become airborne, are transported over a range of distances, and are subsequently deposited, either within the same crop canopy or outside the canopy (Chamberlain and Little, 1981; Waggoner, 1983). Long-distance transport of fungal spores undoubtedly occurs (Hirst et al., 1967; Hermansen, 1982), and procedures for analyzing and predicting such transport, involving techniques such as atmospheric trajectory analysis (Smith, 1983; Pedgley, 1986), offer new possibilities to epidemiologists. Atmospheric trajectory analysis was applied to the case study of tobacco blue mold, *Peronospora tabacina* (Main et al., 1985; Davis et al., 1985; Davis and Main, 1986) to show how a distant inoculum source, identified as south-central Texas, resulted in an explosive epidemic of the disease in North Carolina. Similar techniques have been used to study long-distance transport (Thresh, 1983) of the arthropod vectors of plant virus epidemics (Rosenberg and Magor, 1983). In the first of two chapters concerned with plant virus epidemiology, Irwin and Kampmeier (Chapter 2) consider the myriad ways in which spatial increase of virus disease is affected by vector behavior and ecology, including movement.

A considerable amount of experimental and theoretical work has been done on the transport of fungal spores within plant canopies. Dispersal gradients were described by Gregory (1968, 1973), and their use in many contexts has been reviewed (McCartney and Fitt, 1985, 1987). A possible basis for the different forms of the various empirical equations describing dispersal gradients has been given by McCartney and Bainbridge (1984). Mathematical models based firmly on physical principles have also been proposed to describe two-dimensional movement and deposition of fungal spores within a plant canopy (Legg and Powell, 1979; Aylor, 1982). These models have been further refined, and stochastic elaborations based on turbulence and windgust characteristics of canopies have been proposed (Legg, 1983; McCartney and Fitt, 1985; Fitt and McCartney, 1986).

Models of spore transport should be capable of explaining all events involved in the transport process and should apply over different temporal and spatial frames. Aylor (1986) has provided such a synthesis: A physical model of spore transport is developed that, together with weather records and trajectory analysis, can be used to develop a climatology of risk of disease spread between geographic regions. This key paper in the aerobiology of plant disease should be read in conjunction with the chapters included in this book, and no attempt is made to provide an equivalent synthesis. Of course, there are many other aspects of epidemics apart from spore dispersal that must be evaluated, and the integration of physical spore tranpsort models into epidemic models remains a problematical exercise (Waggoner, 1983; Fitt and McCartney, 1986; McCartney and Fitt, 1987). Berger and Ferris (Chapter 3) take this integration a stage further and consider how transmission processes and spread parameters can be included in simulations of virus spread, although the movement of vectors is not based on a physical model.

Despite the amount of work on airborne spora, however, it should be remembered that not all fungi are dispered in air: Many of the most important plant pathogens are dispersed within splash droplets, dependent on rain, and this can affect the dispersal gradients and the patterns of disease spread observed. Much less work has been done on splash dispersal (McCartney and Fitt, 1987); much remains to be done. Equally, water can be very important in the dispersal of plant pathogenic bacteria over short and long distances.

Franc et al. (1985) provided evidence that viable bacteria can be dispersed within aerosols from the Pacific coast of the United States across the mountain ranges to eastern Colorado.

Soilborne Dispersal

The means of pathogen dispersal that has been least studied is with or through the soil. It has always been considered that dispersal through soil is relatively unimportant whether by growth or by movement [in the case of zoospores and nematodes (Wallace, 1978; Barker, 1985)]. Howard (1985) explored the various means by which *Verticillium* can spread over local and long distances; these include root contact, wind, water, farm machinery, infected plant products, insects, and seed. Site characteristics and preparation may have a major effect on soil dispersal. In only a few cases has hyphal growth through soil been considered a major component of epidemic increase of soilborne pathogens. For *Phymatotrichum omnivorum,* at least, the potential for such spread seems greater than for many plant pathogenic soil fungi (Jeger et al., 1987; Koch et al., 1987).

The soil is the habitat of many vectors of plant virus disease and Thresh (1985) considers the influence of soil vector–virus relationships on plant virus epidemiology. Adams (Chapter 4) considers very unlikely candidates for ease of dispersal: These are several mycoparasites of immobile sclerotial-forming fungi. Despite this restricted life history, some of these mycoparasites have a very cosmopolitan distribution. Adams (Chapter 4) also raises the question of nonrandom patterns of soilborne pathogens and mycoparasites and the practical considerations of sampling to detect and quantify their presence.

SPREAD OF PLANT DISEASE

Spatial aspects of epidemics are manifested in more ways than simply as a consequence of the movement and dispersal of the biotic agents causing disease. Irrespective of how the agent (spore, mycelium, vector) causing disease disperses, the patterns of disease that result can offer much insight into disease epidemiology, provided that the observations are made on a spatial scale appropriate to the underlying processes. Four aspects of the spread of disease are considered within chapters of this book: disease gradients; description of patterns of disease, usually on one occasion only but sometimes considered in time as well; the expansion of disease in foci, common in some types of disease; and spatiotemporal analysis, in which disease progress in time and space is considered from a statistical or theoretical (population dynamical) perspective.

Disease Gradients

Disease gradients are not equivalent to and need not reflect dispersal gradients. They cannot be used to deduce mechanisms of dispersal, although they often provide useful information with which to commence an investigation. Confusion sometimes exists as to whether a dispersal gradient or a disease gradient is referred to in the literature. Equally, it should be clear that description of a disease gradient by means of a mathematical function, such as the power law, does not carry the same "physical" conviction as one describing spore dispersal gradients.

Disease gradients are often described by reference to one occasion only (e.g., Kable et al., 1980). Some studies have looked at gradients at successive intervals of time (e.g., Berger and Luke, 1979; Danos et al., 1984), but only recently has mathematical description of gradients been refined to incorporate a time element (Jeger, 1983). These authors also recognized the need to incorporate an upper limit to disease, using transformations based on the logistic or Gompertz functions, in descriptions of gradients—again a feature not usually needed with spore dispersal gradients. It should be noted that disease gradients do not provide the sole means of describing spread. Smeltzer and French (1981) monitored sweet fern rust and trapped spores of *Cronartium comptoniae* with distance from an aeciospore source (pine), using the number of uredinia per spore to characterize both spread and loss of spore viability. Gradients have also been described for plant virus diseases (Thresh, 1976). Allen (1978, 1983) used rather different statistical models to describe the probability of virus infection arising at distances from an infected source and used this as a basis for justifying control actions.

Spatial Patterns of Disease

A contrasting approach to the description of disease gradients, but again usually applied on one occasion only, is analysis of spatial patterns of disease. The objective of most analyses is usually to test the hypothesis of nonrandom patterns. This approach can range from relatively simple techniques such as doublet analysis or ordinary runs analysis (Madden et al., 1982) to very sophisticated techniques described by Madden (Chapter 5). A few studies have combined description of both gradients and spatial pattern (Gitaitis et al., 1978); not surprisingly, the presence of a gradient implies a nonrandom aggregation of diseased plants. Techniques for spatial analysis owe much to pioneering studies in plant ecology and insect populations (see Taylor, 1984) and more recently by human geographers (see Madden, Chapter 5). An aim of such spatial analysis should not only be to obtain good statistical descriptions, but also to obtain insight into the mechanisms that generated the patterns (e.g., the relationship between pathogen and disease spatial patterns) (Schuh et al.,

1986). Recently, some studies have looked at spatial patterns as they change over time and in relation to temporal models of increase (Madden et al., 1987); there is now the prospect of fully integrated spatiotemporal models that combine both spatial pattern and disease progress with time.

Disease Foci and Spread

Before spatiotemporal models are discussed, the case of disease gradients will be examined again as the starting point for modeling the growth of disease foci (Jeger, 1985). Disease foci and their expansion are a characteristic feature of many explosive epidemics: for example, the cereal rusts (Zadoks and Schein, 1979; Kingsolver et al., 1984). Empirical models (Jeger, 1983) provide a description of focus expansion, but more considered models have since been proposed combining both simplicity and elements of biological realism. Minogue (Chapter 6) summarizes previous work (see also Minogue, 1986) and considers models, termed spatial probability models, that lead to characteristic features of epidemics, such as wavelike spread in which there is a constant velocity of outward spread from the focus center. This type of wavelike spread arises in many contexts and has been discussed for predator–prey interactions (Dunbar, 1983) and in the context of human epidemics (Bailey 1975). For plant disease epidemics the analysis has been extended by workers in the Netherlands (Diekmann, 1986; Heesterbeek and Zadoks, 1987; Bosch et al., 1987a,b). The linking of these models to experimental observations represents an advance in the understanding of mechanisms by which focal diseases progress. It should also be noted that descriptions of spread other than in foci can be found (see Chapter 6) and many innovative ways of describing spread of nonpathogenic events in crops (Fujioka 1985) may have application for plant diseases.

Spatiotemporal Analysis

In discussing spatial spread of plant disease, several aspects introduced above can now be integrated to anticipate future developments in spatiotemporal analysis.

Statistical approaches. From the statistical approach directed to spatial patterns, techniques have now been developed which permit analysis of epidemics using spatiotemporal autocorrelation. Madden (Chapter 5) discusses the potential and limitations of this sort of analysis, which is still in its infancy. In general the descriptions of epidemics that are obtained by spatiotemporal autocorrelation are based on empirical functions that are lagged in time and space. It is, perhaps, illustrative of the increasing maturity of plant disease epidemiology, and certainly of the innovation of Madden and coworkers, that some of the first applications have been made in the discipline. There are many extensive data in the literature (e.g., Dale, 1953) that could be evaluated using spatiotemporal analysis, and awareness of the techniques could influence the ways in which data are collected in the future.

Population dynamic approaches. Zadoks and Kampmeijer (1977) used a spatial elaboration of Vanderplank's (1963) differential-difference equation with a spore dispersal model to simulate disease spread on a two-dimensional grid. There is, alternatively, an entire class of theoretical population dynamic models with spatial elements, based largely on diffusion-type processes with or without birth–death processes (see Levin, 1981) that can be used instead of explicit spatial representation. About the only case study in plant diseases is that of Fleming et al. (1982), which was used to explore the effects of field geometry and size on disease dynamics. In pest systems, such models have been more common and have been further elaborated by inclusion of environmental/crop factors in extensive simulation models (Sawyer and Haynes, 1985).

SPATIAL VARIABILITY AND PLANT DISEASE

Many descriptions or analyses of plant disease epidemics include the assumption of uniform conditions, either biotic or abiotic, in time and space. Certainly, mathematical models, whether analytical or simulation (*sensu* Jeger, 1986), usually ignore the spatial variability that is present in all agricultural systems, although due deference may be given to time-varying environments and thus parameters. In natural ecosystems the importance of spatial variability, whether genetic or environmental, has long been recognized; Alexander (Chapter 7) offers a considered synthesis of the various forms of heterogeneity and their effects on plant disease epidemics. A unique feature is the emphasis placed on the manipulative field experiments that are necessary to test hypotheses concerning spatial effects in epidemiology. The importance of spatial heterogeneity in agricultural systems can be realized by considering the effectiveness of genetic heterogeneity in multilines or cultivar mixtures (Wolfe, 1985) as a specific measure to control plant diseases. The role of environmental diversity in crop protection in general has been reviewed by Perrin (1980).

Groth and Alexander (Chapter 8) continue with the theme of spatial heterogeneity by considering genetic divergence and diversity of pathogen populations that reflect heterogeneity and influence the course of epidemics. Both phenomena occur on scales ranging from a few centimeters to many kilometers, and there is imperfect knowledge of how divergence and diversity are generated and maintained over this range of scales. There is also an obverse side to diversity which is relevant for plant disease epidemics: Often the occurrence of a pathogenic strain in a given locality can be used to infer where it originated, although most reports of this have come from entomological studies (Pashley, 1985).

The management of host diversity and heterogeneity is considered by Marshall (Chapter 9) in the context of resistance breeding and germplasm demployment. A contrast is made between germplasm deployment as a solution to disease problems and the new problems that this deployment may create. It is clear that each solution carries within it the potential for a new problem; that is why the time frame of effective research in plant disease epidemiology, especially in research at the pathology/breeding interface, should not be considered or appraised in terms of only a few years. Plant breeders must also be concerned with the yield of crops; in developing breeding programmes plant diseases may represent only one of the constraints. There have been numerous studies of the effects of disease on crop yield. Methodologies for developing and appraising crop loss–disease relationships have been proposed. Yet virtually none of these studies consider the inherent variability in crop yield, often expressed spatially, and the impact of this variability on estimates of loss caused by plant disease. This deficiency is equally present in the work of crop modelers who develop models reflecting the physiological mechanisms within individual plants without considering crops as populations of plants and the variation encountered within populations. Gaunt and Robertson (Chapter 10) go some way to redressing the balance and set a realistic framework for those who wish to derive disease–crop loss models.

In the final chapter of this book, Appel (Chapter 11) considers disease in an environment that is uniquely heterogeneous: the urban forest. Many forest ecosystems, especially those that are natural or quasi-natural, offer unique opportunites for studying epidemics as affected by spatial heterogeneity. The effects of forest pathogens can be considered on many scales, from the continental (Gibbs and Wainhouse, 1985) to the national or provincial (Gibbs, 1978; Kurkela, 1984) to the local (Cicinsky, 1978; Menges and Kutz, 1985). The spatial aspects of forest epidemics are often self-apparent; the urban forest provides further insights into the effects of disease epidemics (Karnosky, 1979). Appel (Chapter 11) considers the many issues, some as much sociological as silvicultural, that are raised by this environment, entirely human-centered, yet possessing few of the features of agricultural monocultures.

Recently, the view was expressed that the dynamics of disease in spatially heterogeneous populations is one of the major outstanding problems in plant disease epidemiology (Jeger, 1986). The succeeding chapters in this book amply describe the nature of the problem, at the theoretical and practical ends of the spectrum of approaches in plant disease epidemiology. The questions that arise and the solutions that must be sought will command the attention of epidemiologists and plant pathologists, in general, for the foreseeable future.

REFERENCES

ALLEN, R. N. 1978. Spread of bunchy top disease in established banana plantations. Aust. J. Agric. Res. 29:1223–1233.

ALLEN, R. N. 1983. Spread of banana bunchy top in time and space. *In* Plant Virus Epidemiology: The Spread and Control of Insect-Borne Virus (ed. R. T. Plumb and J. M. Thresh), pp. 51–59. Blackwell Scientific Publications, Ltd., Oxford.

AYLOR, D. E. 1982. Modelling spore dispersal in a barley crop. Agric. Meteorol. 26:215–219.

AYLOR, D. E. 1986. A framework for examining inter-regional aerial transport of fungal spores. Agric. For. Meteorol. 38:263–288.

BAILEY, N. T. J. 1975. The Mathematical Theory of Infectious Diseases and Its Applications. Charles Griffin & Company Ltd., High Wycombe, Buckinghamshire, England.

BARKER, K. R. 1985. A history of the introduction and spread of nematodes. *In* The Movement and Dispersal of Agriculturally Important Biotic Agents (ed. D. R. MacKenzie, C. S. Barfield, G. C. Kennedy, and R. D. Berger), pp. 131–144. Claitor's Publishing Division, Baton Rouge, La.

BERGER, R. D., and LUKE, H. H. 1979. Spatial and temporal spread of oat crown rust. Phytopathology 69:1199–1201.

BOSCH, F. VAN DEN, ZADOKS, J. C., and METZ, J. A. J. 1988a. Focus formation in plant disease. I. The constant rate of focus expansion. Phytopathology 78:54–58.

BOSCH, F. VAN DEN, ZADOKS, J. C., and METZ, J. A. J. 1988b. Focus formation in plant disease. II. Realistic parameter-sparse models. Phytopathology 78:59–64.

CHAMBERLAIN, A. C., and LITTLE, P. 1981. Transport and capture of particles by vegetation. *In* Airborne Dispersion (21st Symposium of the British Ecological Society), (ed. J. Grace, E. D. Ford, and P. G. Jarvis), pp. 147–173. Blackwell Scientific Publications, Ltd., Oxford.

CICINSKY, M. J. 1978. The diffusion of dutch elm disease in Warwickshire. Unpublished B.Sc. project, Department of Geography, Lanchester Polytechnic.

DALE, W. T. 1953. Further notes on the spread of virus in a field of clonal cacao in Trinidad. *In* Cocoa Research 1945–1951. pp. 130–131. Imperial College of Tropical Agriculture, Trinidad.

DANOS, E., BERGER, R. D., and STALL, R. E. 1984. Temporal and spatial spread of citrus canker within groves. Phytopathology 74:904–908.

DAVIS, J. M., and MAIN, C. E. 1986. Applying atmospheric trajectory analysis to problems in epidemiology. Plant Dis. 70:490–497.

DAVIS, J. M., MAIN, C. E., and NESMITH, W. C. 1985. The biometeorology of blue mould of tobacco. II. The evidence for long-range sporangiospore transport. *In* Movement and Dispersal of Agriculturally Important Biotic Agents (ed. D. R. MacKenzie, C. S. Barfield, G. C. Kennedy, and R. D. Berger), pp. 471–496. Claitor's Publishing Division, Baton Rouge, La.

DIEKMANN, O. 1986. Dynamics in bio-mathematical perspective. *In* Mathematics and Computer Science II (ed. M. Hazelwinkel et al.), pp. 23–50. North-Holland Publishing Company, Amsterdam.

DUNBAR, S. R. 1983. Travelling wave solutions of diffusive Lotka–Volterra equations. J. Math. Biol. 17:11–32.

FITT, B. D. L., and MCCARTNEY, H. A. 1986. Spore dispersal in relation to epidemic models. *In* Plant Disease Epidemiology: Population Dynamics and Management, Vol. 1 (ed. K. J. Leonard and W. E. Fry), pp. 311–345. Macmillan Publishing Company, New York.

FLEMING, R. A., MARSH, L. M., and TUCKWELL, H. C. 1982. Effect of field geometry on the spread of crop disease. Prot. Ecol. 4:81–108.

FRANC, G. D., HARRISON, M. D., and POWELSON, M. L. 1985. The dispersal of phytopathogenic bacteria. *In* The Movement and Dispersal of Agriculturally Important Biotic Agents (ed. D. R. MacKenzie, C. S. Barfield, G. C. Kennedy, and R. D. Berger), pp. 37–49, Claitor's Publishing Division, Baton Rouge, La.

FRY, W. E. 1982. Principles of Plant Disease Management. Academic Press, Inc., New York.

FUJIOKA, F. M. 1985. Estimating wildland fire rate of spread in a spatially nonuniform environment. For. Sci. 31:21–29.

GIBBS, J. N. 1978. Development of the Dutch elm disease epidemic in southern England, 1971–6. Ann. Appl. Biol. 88:219–228.

GIBBS, J. N., and WAINHOUSE, D. 1985. The spread of forest pests and pathogens in the northern hemisphere. *In* The Movement and Dispersal of Agriculturally Important Biotic Agents (ed. D. R. Mackenzie, C. S. Barfield, G. C. Kennedy, and R. D. Berger) pp. 265–277. Claitor's Publishing Division, Baton Rouge, La.

GILLIGAN, C. A. (ed.) 1985. Advances in Plant Pathology, Vol. 4, Mathematical Modelling of Crop Diseases. Academic Press, Inc. (London) Ltd., London.

GITAITIS, R. D., STALL, R. E., and STRANDBERG, J. O. 1978. Dissemination and survival of *Pseudomonas alboprecipitans* ascertained by disease distribution. Phytopathology. 68:227–231.

GREGORY, P. H. 1968. Interpreting plant disease dispersal gradients. Annu. Rev. Phytopathol. 6:189–212.

GREGORY, P. H. 1973. The Microbiology of the Atmosphere. Leonard Hill, London.

HEESTERBEEK, J. A. P., and ZADOKS, J. C. 1987. Modelling pandemics of quarantine pests and diseases: problems and perspectives. Crop Prot. 6:211–221.

HERMANSEN, J. E. 1982. Wind dispersal of spores of cereal pathogens across the North Sea. Garcia de Orta Ser. Estud. Agron. (Lisb.) 9(1–2):143–146.

HIRST, J. M., STEDMAN, O. J., and HURST, G. W. 1967. Long-distance spore transport: vertical sections of spore clouds over the sea. J. Gen. Microbiol. 48:357–377.

HORSFALL, J. G., and COWLING, E. A. (eds.) 1978. Plant Disease: An Advanced Treatise, Vol. 2, How Disease Develops in Populations. Academic Press, Inc. New York.

HOWARD, R. J. 1985. Local and long-distance spread of *Verticillium* species causing wilt of alfalfa. Can. J. Plant Pathol. 7:199–202.

JEGER, M. J. 1983. Analysing epidemics in time and space. Plant Pathol. 32:5–11.

JEGER, M. J. 1985. Models of focus expansion in disease epidemics. *In* The Movement and Dispersal of Agriculturally Important Biotic Agents (ed. D. R. MacKenzie, C. S. Barfield, G. C. Kennedy, and R. D. Berger), pp. 279–288. Claitor's Publishing Division, Baton Rouge, La.

JEGER, M. J. 1986. The potential of analytic compared with simulation approaches in plant disease epidemiology. *In* Plant Disease Epidemiology: Population Dynamics and Management, Vol. 1 (ed. K. J. Leonard and W. E. Fry), pp. 255–281. Macmillan Publishing Company, New York.

JEGER, M. J., KENERLEY, C. M., GERIK, T. J., and KOCH, D. O. 1987. Spatial dynamics of Phymatotrichum root rot in row crops. Phytopathology 77:1647–1656.

KABLE, P. F., FRIEND, P. M., and MACKENZIE, D. R. 1980. The spread of a powdery mildew of peach. Phytopathology 70:601–604.

KARNOSKY, D. F. 1979. Dutch elm disease: a review of the history, environmental implications, control, and research needs. Environ. Conserv. 6:311–322.

KINGSOLVER, C. H., PEET, C. E., and UNDERWOOD, J. F. 1984. Measurement of the epidemiological potential of wheat stem rust: St. Croix, U.S. Virgin Islands, 1954–1957. Pa. Agric. Exp. Stn. Bull. 854.

KOCH, D. O., JEGER, M. J., GERIK, R. J., and KENERLEY, C. M. 1988. The effect of plant density on the progress of *Phymatotrichum* root rot in cotton. Phytopathology 77:1657–1662.

KRANZ, J. (ed.) 1974. Epidemics of Plant Diseases: Mathematical Analysis and Modelling, Springer-Verlag, Berlin.

KURKELA, T. 1984. Factors affecting the development of disease epidemics by *Gremmeniella abietina*. For. Sci. 13:148–152.

LEGG, B. J. 1983. Movement of plant pathogens in the crop canopy. Philos. Trans. R. Soc. London Ser. B 302:559–574.

LEGG, B. J., and POWELL, F. A. 1979. Spore dispersal in a barley crop: a mathematical model. Agric. Meteorol. 20:47–67.

LEONARD, K. J., and FRY, W. E. (eds.) 1986. Plant Disease Epidemiology: Population Dynamics and Management, Vol. 1. Macmillan Publishing Company, New York.

LEVIN, S. 1981. Models of population dispersal. *In* Differential Equations and Applications in Ecology, Epidemics, and Population Problems (ed. S. N. Busenberg and K. L. Cooke), pp. 1–18. Academic Press, Inc., New York.

MACKENZIE, D. R., BARFIELD, C. S., KENNEDY, G. C., and BERGER, R. D. (eds.) 1985. The Movement and Dispersal of Agriculturally Important Biotic Agents. Claitor's Publishing Division, Baton Rouge, La.

MADDEN, L. V., LOUIE R., ABT, J. J., and KNOKE, J. K. 1982. Evaluation of tests for randomness of infected plants. Phytopathology 72:195–198.

MAIN, C. E., DAVIS, J. M., and MOSS, M. A. 1985. The biometeorology of blue mould of tobacco, Part I: A case study of the epidemiology of the disease. *In* The Movement and Dispersal of Agriculturally Important Biotic Agents (ed. D. R. MacKenzie, C. S. Barfield, G. C. Kennedy, and R. D. Berger), pp. 453–471. Claitor's Publishing Division, Baton Rouge, La.

McCartney, H. A., and Bainbridge, A. 1984. Deposition gradients near to a point source in a barley crop. Phytopath. Z. 109:219–236.

McCartney, H. A., and Fitt, B. D. L. 1985. Construction of dispersal models. *In* Advances in Plant Pathology, Vol. 4, Mathematical Modelling of Crop Disease (ed. C. A. Gilligan), pp. 107–143. Academic Press, Inc. (London) Ltd., London.

McCartney, H. A., and Fitt, B. D. L. 1987. Spore dispersal gradients and disease development. *In* Populations of Plant Pathogens: Their Dynamics and Genetics (ed. M. S. Wolfe and C. E. Caton), pp. 109–118. Blackwell Scientific Publications Ltd., Oxford.

Menges, E. S., and Kutz, J. E. 1985. Predictive equations for local spread of oak wilt in southern Wisconsin. For. Sci. 31:43–51.

Minogue, K. P. 1986. Disease gradients and the spread of disease. *In* Plant Disease Epidemiology: Population Dynamics and Management, Vol. 1 (ed. K. J. Leonard and W. E. Fry), pp. 285–310. Macmillan Publishing Company, New York.

Pashley, D. P. 1985. The use of population genetics in migration studies: a comparison of three noctuid species. *In* The Movement and Dispersal of Agriculturally Important Biotic Agents (ed. D. R. MacKenzie, C. S. Barfield, G. C. Kennedy, and R. D. Berger), pp. 305–324. Claitor's Publishing Division, Baton Rouge, La.

Pedgley, D. E. 1986. Long distance transport of spores. *In* Plant Disease Epidemiology: Population Dynamics and Management, Vol. 1 (ed. K. J. Leonard and W. E. Fry), pp. 346–365. Macmillan Publishing Company, New York.

Perrin, R. M. 1980. The role of environmental diversity in crop protection. Prot. Ecol. 2:77–114.

Plumb, R. T., and Thresh, J. M. (eds.) 1983. Plant Virus Epidemiology: The Spread and Control of Insect-Borne Viruses. Blackwell Scientific Publications Ltd., Oxford.

Rosenberg, J. L., and Magor, J. I. 1983. A technique for examining the long-distance spread of plant virus diseases transmitted by the brown planthopper *Nilaparvata lugens* (Homoptera: Delphacidae) and other wind-borne insect vectors. *In* Plant Virus Epidemiology: The Spread and Control of Insect-Borne Virus (ed. R. T. Plumb and J. M. Thresh), pp. 229–238. Blackwell Scientific Publications Ltd., Oxford.

Sawyer, A. J., and Haynes, D. L. 1985. Simulating the spatiotemporal dynamics of the cereal leaf beetle in a regional crop system. Ecol. Model. 30:83–104.

Schuh, W., Frederiksen, R. A., and Jeger, M. J. 1986. Analysis of spatial patterns in sorghum downy mildew with Morisita's index of dispersion. Phytopathology 76:446–450.

Scott, P. R. and Bainbridge, A. (eds.) 1978. Plant Disease Epidemiology. Blackwell Scientific Publications Ltd., Oxford.

Smeltzer, D. L. K., and French, D. W. 1981. Factors affecting spread of *Cronartium comptoniae* on the sweet fern host. Can. J. For. Res. 11:400–408.

Smith, F. B. 1983. Meteorological factors influencing the dispersion of airborne diseases. Philos. Trans. R. Soc. London Ser. B 302:439–450.

Taylor, L. R. 1984. Assessing and interpreting the spatial distributions of insect populations. Annu. Rev. Entomol. 29:321–357.

Thresh, J. M. 1976. Gradients of plant virus diseases. Ann. Appl. Biol. 82:381–406.

THRESH, J. M. 1983. The long range dispersal of plant viruses by arthropod vectors. Philos. Trans. R. Soc. London Ser. B 302:497–528.

THRESH, J. M. 1985. Plant virus dispersal. *In* The Movement and Dispersal of Agriculturally Important Biotic Agents (ed. D. R. MacKenzie, C. S. Barfield, G. C. Kennedy, and R. D. Berger), pp. 51–106. Claitor's Publishing Division, Baton Rouge, La.

VANDERPLANK, J. E. 1963. Plant Diseases: Epidemics and Control. Academic Press, Inc., New York.

WAGGONER, P. E. 1983. The aerial dispersal of the pathogens of plant disease. Philos. Trans. R. Soc. London Ser. B 302:451–462.

WALLACE, H. R. 1978. Dispersal in time and space: soil pathogens. *In* Plant Disease: An Advanced Treatise, Vol. 2, How Disease Develops in Populations (ed. J. G. Horsfall and E. B. Cowling), pp. 181–202. Academic Press, Inc., New York.

WOLFE, M. S. 1985. The current status and prospects of multiline cultivars and variety mixtures for disease resistance. Annu. Rev. Phytopathol. 23:251–273.

WOLFE, M. S., and CATON, C. E. (eds.) 1987. Populations of Plant Pathogens: Their Dynamics and Genetics. Blackwell Scientific Publications Ltd., Oxford.

ZADOKS, J. C., and KAMPMEIJER, P. 1977. The role of crop populations and their deployment, illustrated by means of a simulator EPIMUL 76. Ann. N.Y. Acad. Sci. 289:164–190.

ZADOKS, J. C., and SCHEIN, R. D. 1979. Epidemiology and Plant Disease Management. Oxford University Press, Oxford.

2
Vector Behavior, Environmental Stimuli, and the Dynamics of Plant Virus Epidemics

Michael E. Irwin and Gail E. Kampmeier

INTRODUCTION

The dynamics of most plant disease epidemics in agricultural ecosystems are strongly governed by meterological parameters. Rust epidemics, for instance, depend rather directly on wind speed and direction for dispersal and on humidity and dew point for spores to become airborne and subsequently germinate on host tissue (Kranz, 1974). Such systems require the interaction of three components: the host plant of the pathogen and the pathogen, both interacting with the environment.

Epidemics of plant-pathogenic viruses transmitted by arthropods result from recurring movement of infective vectors through ecosystems, with subsequent infections of plants through the introduction of the pathogen by vectors. Epidemics in such systems require the interaction of four notable components: the host plant of the virus, the arthropod vector, and the plant-pathogenic virus itself, each of which interacts with the environment. All components are essential for spread to occur, except when alternative transmission modes are invoked (e.g., vegetative cuttings).

Note: Illinois Natural History Survey and University of Illinois, 607 E. Peabody, Champaign, IL 61820, US.

This quadrad of components intimates that the phenomenon of spread, while more complicated than that of other nonvectored diseases, is deducible and not overly complex. Closer scrutiny, however, reveals that the complex interactions between these components, including the environment, are sufficient to have perplexed epidemiologists for generations. The relationships between pathogenic viruses and their plant hosts are specific and complex (Horsfall and Cowling, 1980; Matthews, 1981). Relationships between plant-pathogenic viruses and their arthropod vectors are often obligatory and specific, resulting from intricate interactions (Pirone and Megahed, 1966; Pirone, 1981; Gergerich et al., 1986). The relationships between arthropod vectors and the virus's host plants are also complicated because of the multitude of behavioral responses the vector can elicit through slight differences in cues and stimuli from a rich and changing environment. That these responses are often species-specific further restricts our understanding of how epidemics proceed. When all of these relationships, each specific and complex, are viewed as parts of a single phenomenon, it becomes clear why virus spread is so difficult to predict.

Environmental stimuli elicit responses that invariably modify vector behaviors (Kennedy et al., 1961; Kring, 1967) which, in turn, alter the pattern and rate of spread of virus epidemics in crop systems. Modification of vector behavior and its effects on the spread of virus epidemics in manipulated crop environments is the topic of this chapter. To investigate this we employ, as a model, a system that we have experimented with over the past decade: soybean mosaic virus.

THE SOYBEAN MOSAIC VIRUS SYSTEM: A MODEL

Soybean mosaic virus (SMV) belongs to the large and economically important potyvirus group whose virus particles are flexuous rods (Gibbs and Harrison, 1976). Its physical and chemical properties are typical of that group (Irwin and Goodman, 1981). SMV presents a constraint to soybean production in parts of the world where early season infections lead to major reductions in seed quantity and quality (Irwin and Goodman, 1981). It will become important in other areas if factors leading to early season spread of the virus become more favorable.

One of the least promiscuous members of the potyvirus group, SMV has a narrow natural host range that is, for practical epidemiological purposes, confined to the family Fabaceae, and, within that, to species of the genus *Glycine* and some close relatives (Galvez, 1963; Irwin and Schultz, 1981).

Based on their virulence reactions to resistant soybean cultivars, seven strains of SMV have been recognized (Cho and Goodman, 1979). All commercially available cultivars of soybean in use in the United States are susceptible to one or more of these currently recognized SMV strains, but five lines developed in Korea appear immune to all seven known strains (Cho and Goodman, 1982).

Yield losses due to SMV have been measured experimentally (Ruesink and Irwin, 1986). As with most potyviruses, yield losses are greatest when the plant becomes infected early. Early infection also maximizes the percentage of seed transmission (Irwin and Goodman, 1981). Seed transmission is perhaps the most important single factor in the dispersal of SMV. Because soybean seed is shipped long distances over relatively short intervals, SMV can be found wherever soybeans are grown. Infected seed is thus the major contributing factor in SMV being the most widespread and economically important of the viruses in soybean.

Seed transmission also accounts for the carryover of virus from one season to the next. Seed for next season's planting is harvested and placed in bins (as seed lots) for storage. When acquiring seed for planting, a grower usually purchases it from a single seed lot. A given proportion of the seeds from any seed lot, when planted, will produce a random distribution of SMV-infected seedlings. If growers in the same area purchase seed from different lots, they could have nearby fields with randomized distributions of different proportions of SMV-infected seedlings. Thus a patchiness of initial inoculum among fields is possible, with a random distribution of inoculum within each field.

The only natural spread of SMV during the growing season is through transmission to noninfected plants in a nonpersistent manner (Schultz et al., 1983) by at least 32 known species of aphids (Insecta: Homoptera: Aphididae) (Irwin and Goodman, 1981; Schultz et al., 1985). In central Illinois, as in most areas of the world, soybean is not colonized by aphids. Spread, in these cases, is limited to transmission by transient, alate (winged) aphids meandering among plants through fields.

The timing and extent of aphid vector movement in soybean fields is important to measure because they account for disease progress in time and space. Aphid landing rates are monitored by mosaic green pan traps (Fig. 1) placed just above the soybean canopy (Irwin and Schultz, 1981; Irwin and Ruesink, 1986). Catches from these pan traps have provided reliable measures of aphid activity by species, and this information has been used to drive the intrafield parameters of an SMV progress model developed by Ruesink and Irwin (1986) (Fig. 2).

In this model, the probability of a healthy plant becoming infected during a given 24-h period depends on the number of source plants present, the total number of plants, and the measured landing rate of each aphid species. Because host plants are presumed to be randomly distributed within the field, no spatial component currently exists in this model. The model considers only intrafield buildup of the disease and assumes that neighboring fields have comparable or lower levels of SMV infection. Furthermore, the model assumes that aphids moving within fields spread the virus in proportion to the number of source plants available and that immigrants carry no SMV with them. In

Figure 1 A horizontal mosaic green pan trap is maintained just above the soybean canopy within a designated row of plants. The trap consists of a 12 cm × 12 cm mosaic green ceramic tile (Cambridge 815 from Cambridge Tile Co., P.O. Box 15071, Cincinnati, OH 46215) placed in a plastic sandwich box container (Sterilite Corp., Townsend, MA 01469). The box is filled with a 50% aqueous solution of ethylene glycol and mounted to a metal pole with a double chemistry clamp, so that it may be raised to the level of the soybean canopy. (Photo by R. Korb.)

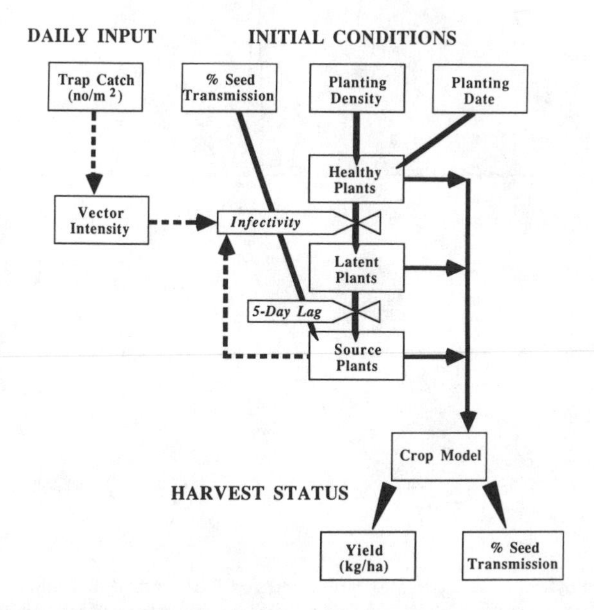

Figure 2 Flow diagram of SMV simulation model developed by W. G. Ruesink with the collaboration of M. E. Irwin and R. M. Goodman. The model predicts disease progress during the growing season and estimates yield loss and percentage seed transmission from the resulting harvest.

other virus–vector systems, such as the persistently transmitted barley yellow dwarf virus (Wallin and Loonan, 1971) and the nonpersistently transmitted maize dwarf mosaic virus (Zeyen et al., 1978), the pathogen may be carried long distances by migrating vectors.

Forecasting the buildup and impact of SMV necessitates a daily knowledge of total aphid landings, their species composition and their vector propensities (i.e., the probability that a single vector specimen, having had the opportunity to acquire a virus in nature by landing on an infected plant, will subsequently transmit the virus, provided that it lands on a noninfected host plant of the virus in nature), which are species specific. A daily measure of vector intensity is thus the sum of the propensity of each species multiplied by its landing rate (Fig. 3A) (Irwin and Ruesink, 1986). Absolute landing rates of any given species combine plant-to-plant movement and long-distance migration. The SMV model could handle both aspects of landing rates, but long-

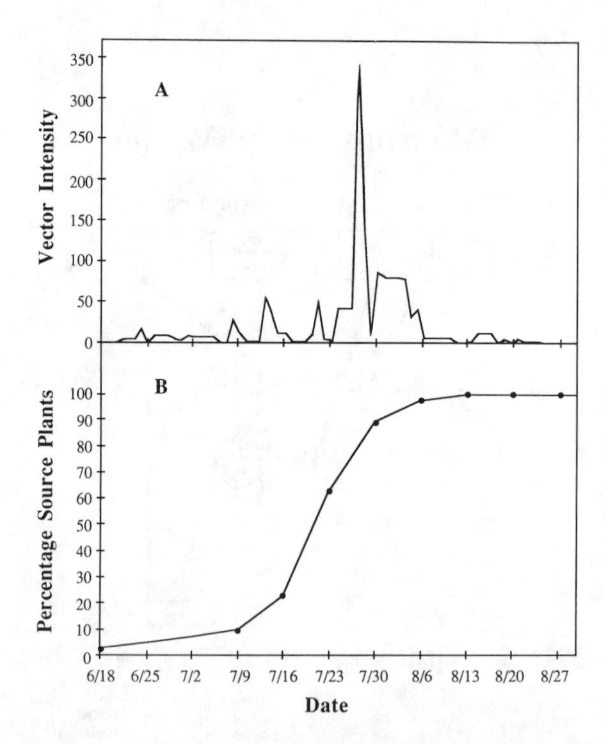

Figure 3 Soybean, cv. Essex was monitored for SMV spread and aphid landing rates in Urbana, Illinois, 1982. Landing rates are transformed into measures of vector intensity (see the text) over the growing season (A). The incidence of SMV was recorded weekly (B).

distance and local movement components of given vector species cannot yet be resolved.

The temporal pattern of intrafield SMV disease progress is reasonably well defined (Fig. 3B). It conforms well to the spread of similar aphid-borne, nonpersistently transmitted viruses in row crops (Swenson, 1968; Thresh, 1974, 1980; Conti et al., 1979). The progress curve is sigmoid, generally steep, and often approaching 100% within the season. The steepness of the curve is highly correlated with the quantity and timing of vector species activity (Schultz et al., 1985).

The spatial pattern of SMV spread from a virus source within a field (Fig. 4) (Irwin and Goodman, 1981) is, in terms of gradient and distance, consistent with many nonpersistently transmitted aphid-borne virus studies (Demski and Sumner, 1979; Jayasena and Randles, 1984); however, spread from the source appears to be more spatially constrained than that reported by Hampton (1967), comprising a steep gradient, as generally discussed by Thresh (1976). This difference could well be due to the fact that, in Hampton's studies, virus entered the crop with the vectors; it was not contained within the field. Because aphids do not colonize soybean in central Illinois and because seed transmission is the major inoculum source, alate aphids, originating outside weed-free soybean fields, are responsible for acquiring and spreading SMV largely within the field.

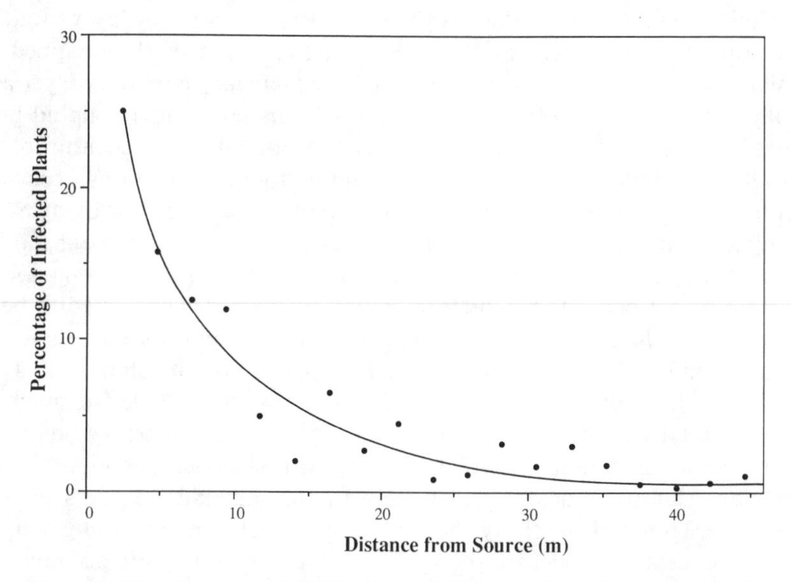

Figure 4 Mean incidence of SMV-infected Williams soybean plants at radial distances from a source of initial inoculum, Tolono, Illinois, 1976.

AERIAL MOVEMENT OF APHID VECTORS: CONCEPTUAL MODELS

The SMV epidemiology model depends on a knowledge of day-to-day aphid catches. This means that the model cannot be predictive until aphid landings can be forecast. A conceptual framework was developed to predict the intrafield spread of SMV; it requires a knowledge of intrafield, interfield, and long-distance movement patterns of the major vector species. We have begun to study these movements by examining one of the major vector species of SMV in central Illinois, the corn leaf aphid, *Rhopalosiphum maidis* (Fitch) (Homoptera: Aphididae) (Halbert et al., 1981).

We are attempting to develop, through a model, the capability to predict the type, quantity, and timing of immigrating vectors by determining source areas, describing vector temporal and spatial distribution in the atmosphere, and studying processes of vector flight initiation, flight termination, and local movements. This model consists of a horizontal translation component and a vertical displacement component (Hendrie et al., 1985).

Vertical Displacement Submodel

Our submodel of vertical displacement simplifies reality by dividing the troposphere into four layers (Fig. 5), the lowest representing the aphid pool on plants. The layer immediately above the crop canopy contains the aphid pool within the surface boundary layer (lower 10 to 20 m). The layer of air in which turbulence and surface effects dominate and that is often capped by inversions during aphid migrations is called the planetary boundary layer and typically is about 1 km deep. The uppermost layer represents the aphid pool that has become involuntarily uplifted by convection into the free atmosphere above the planetary boundary layer. Maximum aphid interactions occur between the pool of landed aphids, the most important layer for studying SMV epidemiology, and all other layers. Individual aphids, according to our model, voluntarily leave this layer by local takeoff or migration takeoff behavior.

Aphids that occur in the surface boundary layer are considered to be in a local or short-duration movement and will land but will not move into the planetary boundary layer. This pool is responsible for the short, plant-to-plant, intrafield movement patterns and is directly accountable for much of the virus spread within a field. Aphids that occur in the planetary boundary layer represent true migrants. Movement from this layer to the pool of aphids in the plant canopy is determined largely by the individual aphid's physiological state, largely dictated by the depletion of fuel reserves through flight.

The conceptual model for the vertical component takes into account voluntary transport of the aphids during their ascent and descent and involuntary ascent from lower levels when convective currents produce vertical wind veloc-

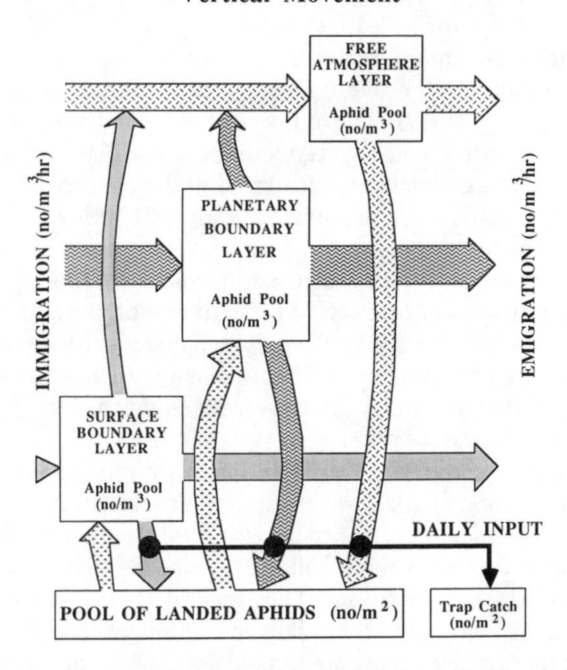

Figure 5 Conceptual model, developed by Hendrie et al. (1985), of vertical displacement of immigrating and emigrating aphids, including dynamic movement between atmospheric layers and landing in plant canopies.

ities in excess of an aphid's maximum flight speed. Meteorological models of air movement are currently used to predict aphid movement.

Horizontal Translocation Submodel

Because aphid flight speed is low compared with that of wind under most conditions, the model assumes that aphid horizontal translocation is controlled by air movement. A case study was conducted on aerial movement of *R. maidis* into central Illinois on August 9, 1984 (Hendrie and Irwin, 1986). In that case study, *R. maidis* specimens were collected between 7 and 8 A.M. in two distinct layers, 450 m and 950 m above ground level. Aphids in the upper and lower layers had an average lipid content of 22.7% and 31.8% of dry body weight, respectively. Using laboratory-reared specimens of *R. maidis,* Liquido and Irwin (1986) demonstrated a negative correlation of lipids with flight duration described by the regression equation, $y = 555.47 - 9.62x + 0.77x^2 - 0.02x^3$. Using the correlation obtained, and assuming that aphids do not take off or land in the dark, it was possible to demonstrate that the aphids in the lower layer had flown for no more than 2 h while those in the upper layer had flown for at least 11 h.

Wind transport, a major factor in the long-distance movement of aphids, can be measured by an objective trajectory model successfully developed by Scott and Achtemeier (1987). This allows for the computation of horizontal translocation of aphids and accurately describes their flight paths. When combined with flight energy analyses, illustrated above, it provides the potential for good resolution of source regions under all meteorological conditions. In the case study above, the aphids in the lower layer originated near Springfield, Illinois, some 100 km to the west of Champaign, while those from the upper layer originated from southwestern Missouri or northeastern Texas. When insect–weather relationships are sufficiently well understood, the objective wind transport model can be converted from a diagnostic to a prognostic mode. Potential flight paths and likely landing areas could then be predicted based on forecasted winds and weather.

A knowledge of the aerial distribution of migrating insects, including their elevation, density, spatial organization, and relationship to meteorological parameters, is crucial to further development of the objective trajectory model. Helicopter-mounted aerial collectors provide reliable data and are usable under most weather conditions. They accurately sample absolute volumes of air, allowing the computation of realistic insect densities, provide the capability to partition samples according to predetermined requirements, and provide undamaged specimens suitable for identification and biological assays.

Preliminary data from central Illinois (Irwin and Hendrie, unpublished) suggest that migrating aphids prefer prefrontal conditions of moderate to strong southwesterly flows of air. Specimens of *R. maidis* are usually found in distinct layers, apparently associated with temperature inversions or wind maxima, or both.

APHID RESPONSES TO ENVIRONMENTAL STIMULI

Aphids in local or migratory flight modes may well respond differently to probing and landing stimuli; however, we have yet to test these potential differences. Probing and landing activities, along with stimuli that influence flight patterns, are very important in the spatial and temporal dynamics of virus epidemics and are discussed below.

Stimuli That Focus on Probing Activity and Subsequent Host Plant Inoculation

Environmental stimuli that alter vector probing activity and the subsequent inoculation potential of host plants can dramatically change rates of viral epidemics in agroecosystems. These stimuli can be chemical or physical.

We have examined three systems that can be placed in this category. The most obvious is altered probing behavior because of physical properties of soybean leaves. Factors such as host plant resistance to the virus and insecticide effects on the vector are also considered here.

Leaf pubescence. Gunasinghe et al. (1988) examined the influence of leaf pubescence on the spread of SMV by observing landing activity and probing behavior of aphids on foliage with different trichome densities. In a laboratory study, *R. maidis, Myzus persicae* (Sulzer), and *Aphis citricola* Van der Goot, all important vectors of SMV, probed on leaves of soybean isolines that differed in trichome density. Greater probing activity was observed on glabrous and sparsely pubescent than on more densely pubescent isolines. Landing rates of alate aphids in soybean fields were not influenced by leaf trichome density, yet the spread of SMV was retarded in plots containing the more densely pubescent isolines (Fig. 6). Increased leaf trichome density apparently retards virus epidemics by reducing probing frequency and total time spent probing. It does not appear to influence the approach and landing behaviors of vectors.

SMV-resistant genotypes. Gunasinghe et al. (1986) examined transmission-related parameters of aphids on two cultivars, Clark soybean isolines that differed in susceptibility to strains of SMV. Clark 63 was susceptible to strains G1-G7 of SMV, while L78-434 was resistant to all but strain G7 (Cho and Goodman, 1979). Laboratory tests of aphid probing behavior using *M. persicae* showed no significant differences in probe duration, number of probes initiated, or time spent not probing. In field plots planted to these

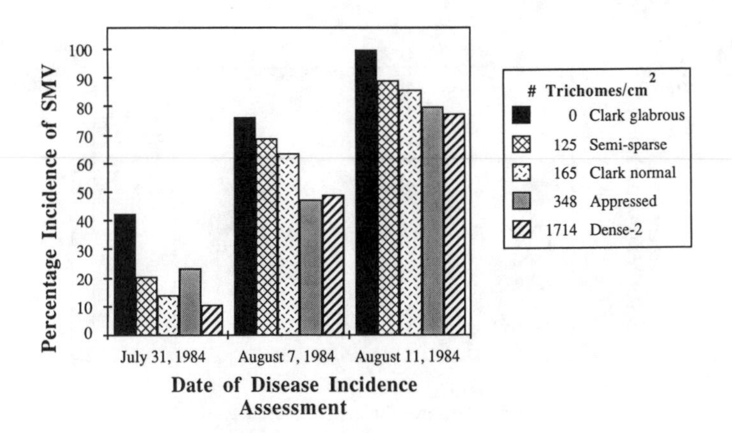

Figure 6 Assessment of SMV incidence in field plots of Clark soybean isolines differing in leaf trichome densities, Urbana, Illinois. (After Gunasinghe et al., 1988.)

two isolines, the incidence of strain G5 increased dramatically, from an initial inoculum level of 2.5%, to 40% infection in Clark 63 but not in L78-434 (Clark 63 seedlings were transplanted into the resistant plots and inoculated to obtain the same initial inoculum). Additionally, no differences in aphid landing rates were detected in plots of these two isolines. The vectors behaved similarly in susceptible and resistant plots and the differences in virus spread can most easily be attributed to the interaction of the virus and its host plant, soybean.

In another field where a susceptible isoline was blended by rows with a resistant isoline, disease incidence was found to be directly proportional to the number of susceptible plants in the blend (Fig. 7) (Irwin and Kampmeier, unpublished). This system involves the vector because the probability of the vector encountering a susceptible plant in different blends can be attributed, not to differences in probing activity, but to different probabilities of encountering susceptible hosts. As suggested by Burdon and Chilvers (1982), the dilution of susceptible plants by those that are nonhosts (i.e., virus resistant) seems a reasonable approach for delaying epidemics long enough to reduce the effects of early infection, such as yield losses, and diminish the threat of seed transmission.

Insecticides. Traditional wisdom would indicate that insecticides cannot prevent the spread of nonpersistently transmitted viruses because they do not act quickly enough (Heathcote, 1973; Gibbs and Harrison, 1976; Quiot et al., 1982). Further, insecticide applications have the potential to increase the spread of such viruses because they increase vector activity (Münster and Murbach, 1952; Broadbent et al., 1963; Shanks and Chapman, 1965).

Quick-acting synthetic pyrethroids, such as clocythrin (Jutsum et al.,

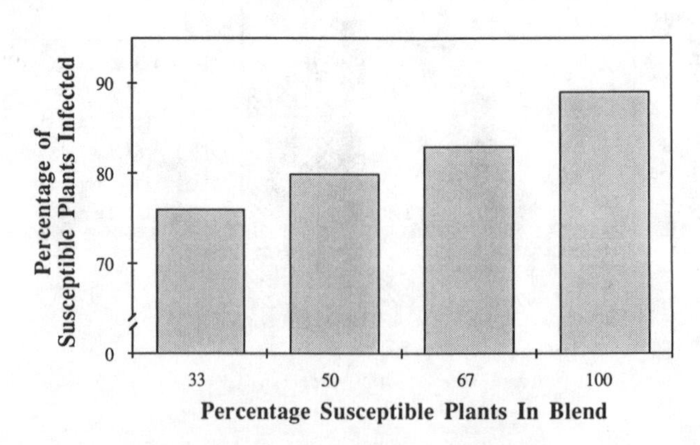

Figure 7 SMV incidence 5 weeks after initial mechanical inoculation of 2.5% of the soybeans with SMV G-5. Plots were planted in mixtures by row of cv. Clark 63 (susceptible) and Clark isoline L78-434 (SMV resistant) in Urbana, Illinois, in 1985.

1984), are purported to prevent epidemics of nonpersistent viruses. An experiment testing this hypothesis was conducted in central Illinois in 1985 (Irwin and Kampmeier, unpublished). The design consisted of four replications of four insecticide treatments: (1) unsprayed control, (2) sprayed control, (3) inoculated row sprayed only, and (4) uninoculated rows sprayed only. The center row of each treatment was mechanically inoculated with SMV and designated plants were sprayed weekly with the insecticide. Infected plants were rogued on a weekly basis to minimize sources for secondary spread. During the critical, preflowering portion of the season, plants showed a reduction in disease incidence in both the sprayed control plots and those where only the inoculated row was sprayed (Fig. 8). Because plots where only the uninoculated rows were sprayed appeared not to retard the spread of the virus significantly relative to that of the unsprayed control, we infer that the insecticide did not act fast enough to prevent inoculation of the virus but could break the cycle of virus spread by preventing acquisition or by incapacitating the aphid before it could move to another plant, or both. Such sprays could be used to retard virus epidemics on crops of high economic value by spraying during a narrow period of crop susceptibility to economic damage.

Stimuli That Focus on Flight Patterns

Many potential stimuli influence the spatial component of aphid flight. Perhaps the most obvious is wind. Wind speed controls takeoff thresholds and, we hypothesize, distance between landings. Wind direction influences di-

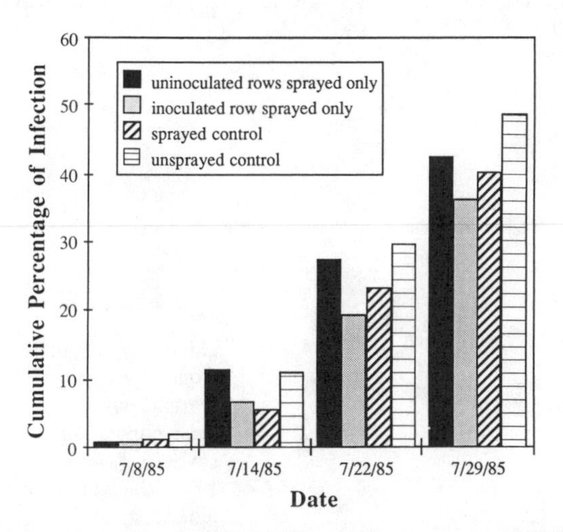

Figure 8 Incidence of SMV at different times during the 1985 growing season in central Illinois in soybean cv. Williams, treated weekly with clocythrin, spread from centrally located rows mechanically inoculated with SMV early in the season.

rection of flight and therefore direction of spread from a virus-infected source plant. Many factors also influence when during the daily cycle an aphid takes flight. Other obvious influences are structural obstacles or barriers. All of these in some way alter the spatial pattern of viral epidemics, and each has been investigated by us.

Wind direction. Aphid flight direction is influenced by wind. The pattern of virus spread from a source generally reflects this influence. In an experiment conducted in central Illinois, the results of this influence can be seen in the pattern of infected plants relative to prevailing winds (Irwin and Goodman, 1981) (Fig. 9). When examined, the slope of the incidence of SMV versus distance from the SMV source is similar for upwind and downwind gradients; however, the magnitude was found to be far greater downwind than upwind (Fig. 10). Although this influence has been noted elsewhere, certain peculiarities in this norm should be mentioned. In an experiment in central Illinois, a wind-directed cylindrical sticky trap monitored aphid flight activity with respect to wind direction in a soybean field. Most species were collected predominantly on the upwind surface of the sticky cylinder; however, one species in particular, *Aphis citricola,* was almost always stuck to the leeward surface. We explain this phenomenon by hypothesizing that *A. citricola* does, indeed, fly upwind but probably does so only when wind velocities are very low (Irwin and Ruesink, 1986).

Another experiment using similar wind-oriented, cylindrical, sticky traps in soybeans demonstrated that some aphid species are more influenced than others by wind direction (Fig. 11). Again, we have little evidence of wind velocity during the time that the specimens were captured, but we hypothesize

Figure 9 Spread (black dots) of SMV from a centralized source of infection (parallel lines; represents 0.5% of soybean population) in a 0.83-ha Williams soybean field near Tolono, Illinois, in 1976. Diagram represents end-of-season disease assessment. Prevailing wind is from the southwest. (Reprinted from Irwin and Goodman, 1981, with permission from Academic Press.)

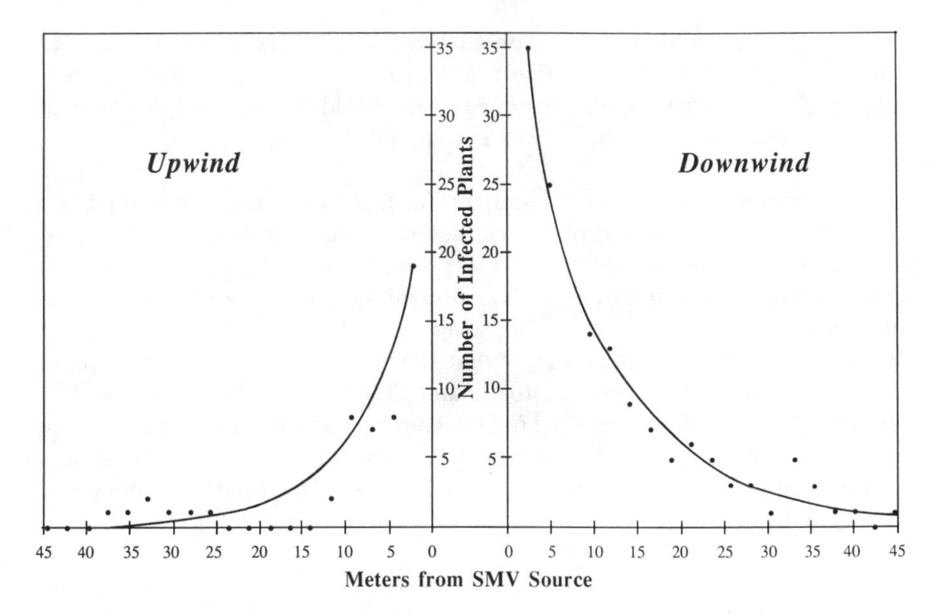

Figure 10 Incidence of SMV infection upwind and downwind of a heavily SMV-infected source area, Tolono, Illinois, 1976.

that part, if not all, of the differences can be accounted for by the velocity of wind during flight times of individual aphid species. We are conducting experiments to test this hypothesis.

Knowledge of such behavior is important when implementing cultural controls designed to modify aphid landing rates, such as barriers or trap crops.

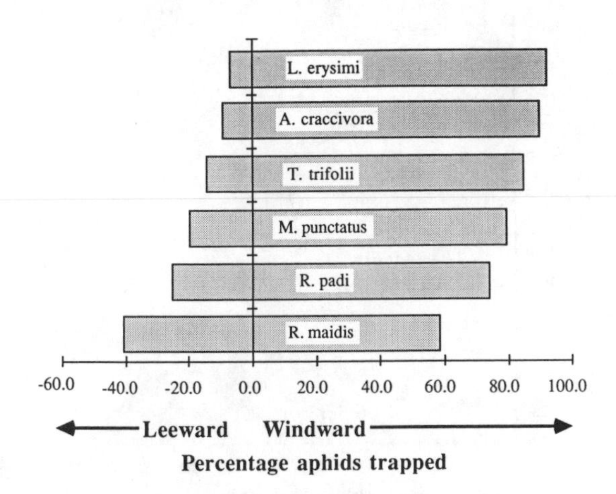

Figure 11 Percentage of aphid species trapped on the leeward and windward surfaces of wind-directed, sticky cylinders maintained at the level of the soybean canopy, Urbana, Illinois, 1978.

Farmers growing more than one crop may wish to consider the prevailing wind when planning where to place fields. For example, small grain fields potentially harboring large populations of vectors should be planted downwind of a soybean field in areas where SMV is a threat.

Diurnal flight cycle and propensity for flight initiation. Although not as important in overall field placement design as wind direction, the timing of diurnal flight of aphid species could well bias the results not only of certain sampling techniques, but of particular control measures as well. Hourly samples from a Johnson–Taylor suction trap (Taylor, 1951), which sifts insects from a known volume of air, show that species differ in their diurnal flight timing (Eastop, 1951; Irwin, Halbert, and Stoetzel, unpublished) (Fig. 12). Sampling techniques that operate only during part of a day may miss key vector species that have different activity cycles (Halbert et al., 1981). Generally, two activity peaks of aphids occur in central Illinois, early in the morning and

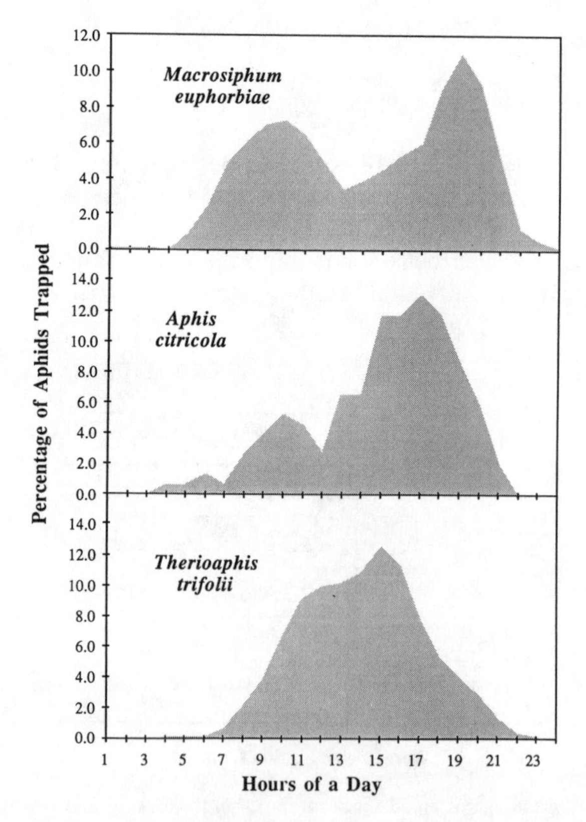

Figure 12 Hourly samples from a Johnson–Taylor suction trap containing three species of aphids flying over soybean, Urbana, Illinois, 1978.

late in the afternoon. Control measures, such as insecticide spraying, that rely on rapid knockdown of aphids will be ineffective if the aphids are not present.

Tied to these diurnal cycles is the propensity to take flight. This may be influenced by such abiotic factors as relative humidity, temperature, and wind speed, and biotic factors such as development (Broadbent, 1949). In wind tunnel experiments with *R. maidis,* for example, tethered aphids would not fly when wind speeds were greater than 2.5 m/s (Zdarkova and Irwin, unpublished). The age of an alate aphid, after becoming an adult, influences its propensity for flight. Alate *R. maidis* that were $\frac{1}{2}$ to 1 day old were most likely to initiate flight at a wind speed of 1 m/s. The propensity to fly by adults less than $\frac{1}{2}$ day old or 2 and 4 days old was greatly decreased, and 8-day-old alates could not be induced to fly under similar conditions (Liquido and Irwin, 1986) (Fig. 13).

Alate aphid flight apparently requires the right combination of aphid physiological and meteorological factors. Knowledge of the effects of meteorological factors at different insect developmental stages would greatly aid in models predicting aphid flight initiation and epidemiology of aphid-borne viruses.

Barriers. Barriers, both living and artificial, have in a number of cases altered the pattern or retarded the timing of plant virus epidemics (Broadbent, 1952, 1957; Jenkinson, 1955; Simons 1957, 1958, 1960; Demski and Kuhn, 1977). Three hypotheses have emerged to explain these results: (1) detention of potential vectors within the barrier, with a subsequent loss in ability to

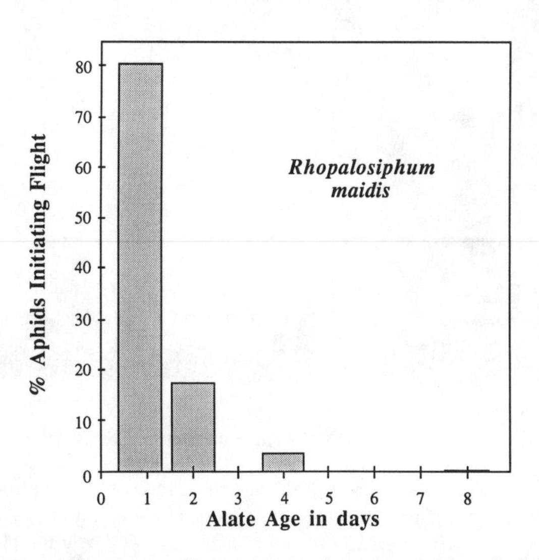

Figure 13 The propensity for flight initiation by tethered *Rhopalosiphum maidis* alate adults aged 0.5 to 8 days in laboratory wind tunnel studies at 1 m/s.

transmit the virus; (2) the physical component of the barrier, which alters the flight path of potential vectors; or (3) both of the above.

Results of our studies investigating the effect of sunflower barriers on the spread of SMV in soybean fields are mixed. Experiments focusing on the pattern and incidence of SMV in plots with and without sunflower barriers strongly indicated that most aphid vectors fly higher in plots with barriers, and therefore, the barriers altered the pattern of SMV in the field (Halbert, 1979). One conclusion from this study was that vectors do not land in or accumulate on the leeward side of barriers. Other studies of ours (Irwin, Halbert, and Schultz, unpublished), however, show variable accumulation of vectors, with resulting higher or lower (Fig. 14) incidence of SMV just downwind of a barrier.

The differences in results of several experiments, we believe, can be explained by barrier porosity, or the ability of air to filter through the barrier. In all cases, we hypothesize, aphids tend to fly over the barrier, but if the barrier is dense, eddies carry the vectors into foliage just downwind of the barrier. If the barrier is quite permeable, airflow is altered, resulting in a system that tends to keep aphids above the canopy for a considerable distance downwind of the barrier. Similar studies by Lewis (1965a,b,c, 1966a,b, 1967, 1969), Jenkinson (1955), Caldwell and Prentice (1942), and Broadbent et al. (1951) suggest the same conclusions.

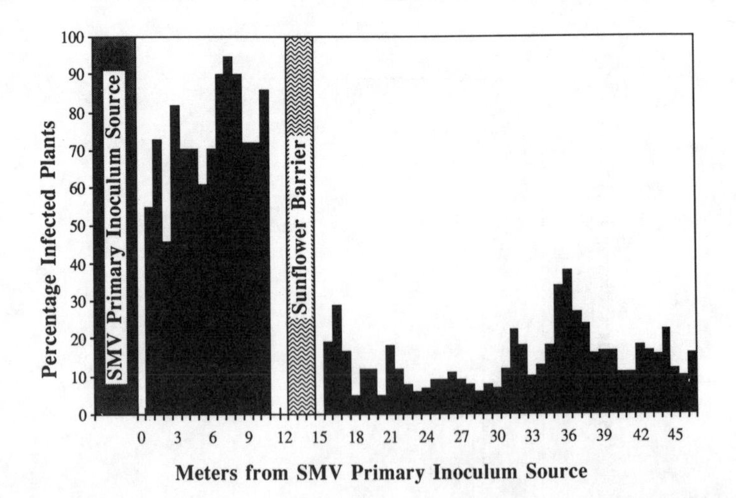

Figure 14 Incidence of SMV both leeward and windward of a sunflower barrier placed downwind of a heavily infected virus source, Urbana, Illinois, 1978 (Halbert, 1979). Source plants were mechanically inoculated with SMV on June 23, 1978, and the graph represents percentage of infected plants on July 11, 1978, when the plants were at growth stage V9 (Fehr et al., 1971).

Stimuli That Focus on Alighting Activity

Certain environmental stimuli elicit alighting responses by vectors. These stimuli can differ in type and degree among vector species. What is perhaps more consequential is that these stimuli can interact in unresolved ways to elicit stronger or weaker responses to different vector species. In our studies thus far, we have investigated canopy cover and foliage color on alighting and subsequent spread of SMV in soybeans.

Canopy cover. We have defined canopy cover simply as the ratio of the amount of ground that is covered by plant material to the amount that is bare. The ratio can be manipulated through cultural (e.g., planting date, planting density, fertilizer, irrigation, and cultivation) and genetic (e.g., cultivar) methods. When monitoring aphids trapped in fields with differing ground cover regimes, researchers (A'Brook, 1964, 1968; Iark and Smith, 1976; Halbert and Irwin, 1981) have found landing rates to be species dependent, with some, such as *Capitophorus elaeagni* (del Guercio), showing a clear preference for sparsely planted fields; others, such as *Lipaphis erysimi* (Kaltenbach), *R. maidis,* and *Schizaphis graminum* (Rondani), showing no preference; and alate *A. citricola* and *Myzocallis punctatus* (Monell) actually preferring a more closed canopy (Halbert and Irwin, 1981).

Researchers who looked at disease spread but not at aphid landing rates have arrived at contradictory or inconclusive results in examining the question of the effect of ground cover (Burdon and Chilvers, 1982). One group (Storey and Ryland, 1957; A'Brook, 1964; Farrell and Adams, 1966, 1967) concluded that sparsely planted fields encouraged more disease, wheras other studies concluded that there were no differences in disease incidence per unit area (Hayes, 1932; Blencowe and Tinsley, 1951; Slykhuis et al., 1960; Booker, 1963; Heathcote, 1969; Kousalya et al., 1971; Davies, 1976).

Our studies suggest that ground cover influences alighting of most aphid vectors of SMV in a similar way: the more ground cover, the fewer landings per unit area. A study in 1979 investigating the effect of planting date on SMV spread concludes that neither plant age nor plant height has any obvious effect on landing rates; however, canopy cover does, not just for some species, but for all species that were trapped during the time of the season when canopy cover differed greatly among treatments (Schultz and Irwin, unpublished) (Fig 15). Another experiment in 1982 investigating the interaction of planting date and row spacing on SMV spread (Irwin and Kampmeier, unpublished) suggests similar conclusions: that canopy cover, not plant age, is the predominant stimulus for aphid alighting.

Foliage color. Aphids are known to be attracted by certain colors (Moericke, 1952, 1955, 1962), especially yellows and greens. A series of replicated field experiments using two isolines of the same cultivar of soybean that

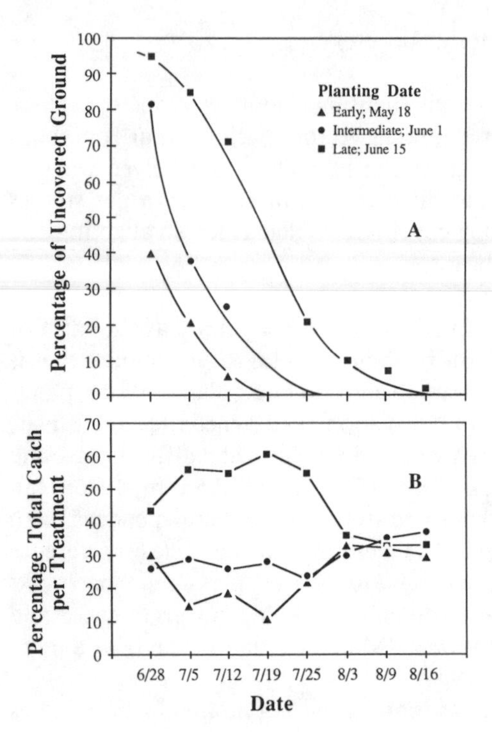

Figure 15 Effect of ground cover on landing by aphids in a soybean field, Urbana, Illinois, 1979: (A) Percentage uncovered ground in three plantings spaced 2 weeks apart; (B) percentage of total aphids trapped in each planting over the season.

differed only in chlorophyll content (Table 1) demonstrated that aphid alighting differed in the two treatments. Alighting ratios were 2:1 in favor of the normal green over the chlorophyll deficient (we perceive this as light yellow) isoline. Not all aphid species reacted to the differences between treatments; however, the resulting decrease in virus spread was about proportional to total aphid landings in the two treatments.

These results may seem inconsistent with those of some other workers, who find that attraction to yellow is often greater than that to green. However,

TABLE 1 Influence of Soybean Leaf Color on Landing Rates of All Alate Aphids onto Clear Glass Tile Traps Treated with Tanglefoot[a] in Central Illinois in 1978

Soybean Leaf Color	Percentage of Aphids Landing	Percentage of SMV Infection
Normal green	65	62
Chlorophyll deficient	35	39

[a]Tanglefoot is a registered trademark of The Tanglefoot Co., 311 Straight Ave. SW, Grand Rapids, MI 49504.

the spectrograph of the two isolines (Fig. 16) clearly demonstrates that the chlorophyll-deficient foliage is not mimicking yellow; it is merely increasing reflectance in part of the green spectrum. This in itself could explain the unexpected results, which were consistent among seasons and over several vector species.

CONCLUDING REMARKS

Most plant viruses, unlike many other plant pathogens, depend heavily on arthropod vectors for spread. Epidemics of plant viruses are governed by many sets of factors, including interactions of the virus with its host plant, the virus with its vector, and the vector with the environment. This last set of factors is particularly important to understand when attempting to model or control plant virus epidemics in agroecosystems. Perhaps the most complex factors involve the interactions of vector behavior with the environment. Subtle changes in the characteristics of the environment can elicit abrupt changes in vector behavior, and this in turn can drastically alter the pattern of virus epidemics in time and space.

In this chapter we have attempted to outline a few of these vector–environmental interactions and how they modify the progress and pattern of soybean mosaic virus epidemics in soybean. Although we have been studying this system for over a decade, we are only beginning to detect the subtlety within which these interactions operate. Continued experimentation with this system, we believe, will result in a greater understanding of vector–environ-

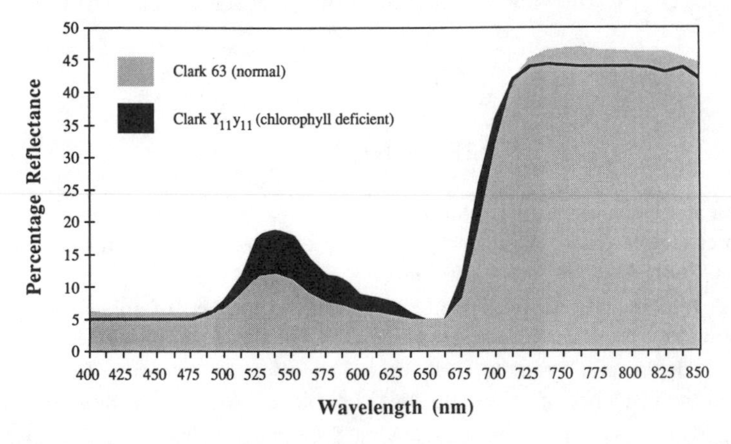

Figure 16 Spectograph of two soybean isolines, cv. Clark 63 and its isogenic line L72-1937 (gene $Y_{11}y_{11}$, lethal semidominant yellow) that is chlorophyll deficient. Doubled young leaves were placed in a Beckman Ratio Recording Spectrophotometer model DK-2A under the direction of J. T. Woolley, Department of Agronomy, University of Illinois at Urbana–Champaign.

ment interactions as a component of plant virus epidemiology. The hypotheses we generate may have wider applicability and can be tested with other systems. A long-term epidemiological goal is to provide a set of principles or guidelines upon which to base virus epidemiological models and control strategies that take into account the total ecology of the virus and its vector in agricultural systems.

ACKNOWLEDGMENTS

This chapter is based on the work of many scientists. They are gratefully acknowledged for their important contributions: Robert M. Goodman, Susan E. Halbert, William G. Ruesink, Gerald A. Schultz, L. Keith Hendrie, Nicanor J. Liquido, U. B. Gunasinghe, Jane E. Polston, and Eva Zdarkova. Many others, too numerous to mention here, also contributed to this program; they too are gratefully acknowledged.

Support for this continuing project has come in part from the Illinois Agricultural Experiment Station (Project 68-0363, Control of Soybean Mosaic Virus Spread through Understanding Virus and Aphid Vectors), from research contract TA/C-1294 from the U.S. Agency for International Development to the International Soybean Program (INTSOY) of the University of Illinois at Urbana–Champaign, through a grant from the Illinois Department of Energy and Natural Resources (STILENPEST-AQ14-752, the Pests and Weather Project), and funds from Imperial Chemical Industries, PLC (Controlling Spread of SMV Using Pyrethroid Insecticides). These sponsors are gratefully acknowledged, but the views expressed in this chapter are those of the authors alone.

REFERENCES

A'BROOK, J. 1964. The effect of planting date and spacing on the incidence of groundnut rosette disease and of the vector, *Aphis craccivora* Koch, at Mokwa, Northern Nigeria. Ann. Appl. Biol. 54:199–208.

A'BROOK, J. 1968. The effect of plant spacing on the numbers of aphids trapped over the groundnut crop. Ann. Appl. Biol. 61:289–294.

BLENCOWE, J. W., and TINSLEY, T. W. 1951. The influence of density of plant populations on the incidence of yellows in sugar beet crops. Ann. Appl. Biol. 38:395–401.

BOOKER, R. H. 1963. The effect of sowing date and spacing on rosette disease of groundnut in northern Nigeria with observations on the vector, *Aphis craccivora.* Ann. Appl. Biol. 52:125–131.

BROADBENT, L. 1949. Factors affecting the activity of alatae of aphids *Myzus persicae* (Sulzer) and *Brevicoryne brassicae* (L.). Ann. Appl. Biol. 36:40–62.

BROADBENT, L. 1952. Barrier crop may help to reduce cauliflower mosaic. Grower 38:1140–1142.

BROADBENT, L. 1957. Investigations of Virus Diseases of *Brassica* Crops. Agr. Res. Coun. Rep. Ser. 14. Cambridge University Press, Cambridge.

BROADBENT, L., TINSLEY, T. W., BUDDIN, W., and ROBERTS, E. T. 1951. The spread of lettuce mosaic in the field. Ann. Appl. Biol. 38:689–706.

BROADBENT, L., GREEN, D. E., and WALKER, P. 1963. Narcissus virus diseases. Daffodil Tulip Yearbook 28:154–160.

BURDON, J. J., and CHILVERS, G. A. 1982. Host density as a factor in plant disease ecology. Annu. Rev. Phytopathol. 20:143–166.

CALDWELL, J., and PRENTICE, I. W. 1942. The spread and effect of broccoli mosaic in the field. Ann. Appl. Biol. 29:374–379.

CHO, E. K., and GOODMAN, R. M. 1979. Strains of soybean mosaic virus: classification based on virulence in resistant soybean cultivars. Phytopathology 69:467–470.

CHO, E. K., and GOODMAN, R. M. 1982. Evaluation of resistance in soybeans to soybean mosaic virus strains. Crop Sci. 22:1133–1137.

CONTI, M., CACIAGLI, P., and CASETTA, A. 1979. Infection sources and aphid vectors in relation to the spread of cucumber mosaic virus in pepper crops. Phytopathol. Mediterr. 18:123–128.

DAVIES, J. C. 1976. The incidence of rosette disease in groundnut in relation to plant density and its effect on yield. Ann. Appl. Biol. 82:489–501.

DEMSKI, J. W., and KUHN, C. W. 1977. A soybean disease caused by peanut mottle virus. Univ. Ga. Agric. Exp. Stn. Res. Bull. 196.

DEMSKI, J. W., and SUMNER, D. R. 1979. Spread of watermelon mosaic virus in Georgia grown squash. Univ. Ga. Agric. Exp. Stn. Res. Bull. 234.

EASTOP, V. F. 1951. Diurnal variation in the aerial density of Aphididae. Proc. R. Entomol. Soc. London Ser. A 26:129–134.

FARRELL, J. A. K., and ADAMS, A. N. 1966. The effect of crop density and time of planting on *Aphis craccivora* Koch. Populations and rosette incidence in groundnuts. Annual Report of the Agricultural Research Council of Central Africa, 1965, pp. 43–49.

FARRELL, J. A. K., and ADAMS, A. N. 1967. The effect of crop density and time of planting on *Aphis craccivora* Koch. Populations and rosette incidence in groundnuts. Annual Report of the Agricultural Research Council of Central Africa, 1966, pp. 55–59.

FEHR, W. R., CAVINESS, C. E., BURMOOD, D. T., and PENNINGTON, J. S. 1971. Stage of development descriptions for soybean, *Glycine max* (L.) Merr. Crop Sci. 11:929–931.

GALVEZ, G. E. 1963. Host-range, purification, and electron microscopy of soybean mosaic virus. Phytopathology 53:388–393.

GERGERICH, R. C., SCOTT, H. A., and FULTON, J. P. 1986. Evidence that ribonuclease in beetle regurgitant determines the transmission of plant viruses. J. Gen. Virol. 67:367–370.

GIBBS, A. J., and HARRISON, B. D. 1976. Plant Virology: The Principles. John Wiley & Sons, Inc., New York.

GUNASINGHE, U. B., IRWIN, M. E., and BERNARD, R. L. 1986. Effect of a soybean genotype resistant to soybean mosaic virus on transmission-related behavior of aphid vectors. Plant Dis. 70:872–874.

GUNASINGHE, U. B., IRWIN, M. E., and KAMPMEIER, G. E. 1988. Soybean leaf pubescence affects aphid vector transmission and field spread of soybean mosaic virus. Ann. Appl. Biol. 112:259–272.

HALBERT, S. E. 1979. Aphid transmission of soybean mosaic virus: some potential means of cultural control. Ph.D. thesis, University of Illinois at Urbana–Champaign.

HALBERT, S. E., and IRWIN, M. E. 1981. Effect of soybean canopy closure on landing rates of aphids with implications for restricting spread of soybean mosaic virus. Ann. Appl. Biol. 98:15–19.

HALBERT, S. E., IRWIN, M. E., and GOODMAN, R. M. 1981. Alate aphid (Homoptera: Aphididae) species and their relative importance as field vectors of soybean mosaic virus. Ann. Appl. Biol. 97:1–9.

HAMPTON, R. O. 1967. Natural spread of viruses infectious to beans. Phytopathology 57:476–481.

Hayes, R. T. 1932. Groundnut rosette disease in the Gambia. Trop. Agric. (Trinidad) 9:211–217.

HEATHCOTE, G. D. 1969. Cultural factors affecting colonization of sugar beet by different aphid species. Ann. Appl. Biol. 63:330–331.

HEATHCOTE, G. D. 1973. Control of viruses spread by invertebrates to plants. *In* Viruses and Invertebrates (ed. A. J. Gibbs), pp. 587–609. North-Holland Publishing Company, Amsterdam.

HENDRIE, L. K., and IRWIN, M. E. 1986. Case study of insect migration. *In* The Pests and Weather Project. (Prep. Illinois State Water Survey and Illinois Natural History Survey), pp. 210–217, ILENR/RE-AQ-87/01.

HENDRIE, L. K., IRWIN, M. E., LIQUIDO, N. J., RUESINK, W. G., MUELLER, E. A., VOEGTLIN, D. J., ACHTEMEIER, G. L., STEINER, W. M., and SCOTT, R. W. 1985. Conceptual approach to modeling aphid migration. *In* The Movement and Dispersal of Agriculturally Important Biotic Agents (ed. D. R. MacKenzie, C. S. Barfield, G. C. Kennedy, and R. D. Berger), pp. 541–582. Claitor's Publishing Division, Baton Rouge, La.

HORSFALL, J. G., and COWLING, E. B. (Eds.) 1980. Plant Disease: An Advanced Treatise, Vol. 5, How Plants Defend Themselves. Academic Press, Inc., New York.

IARK, F., and SMITH, J. C. 1976. Efeito dos espaçamentos do tomateiros ao ataque do *Macrosiphum euphorbiae* (Thomas, 1978) (Homoptera, Aphididae). An. Soc. Entomol. Bras. 5:152–156.

IRWIN, M. E., and GOODMAN, R. M. 1981. Ecology and control of soybean mosaic virus. *In* Plant Diseases and Vectors: Ecology and Epidemiology (ed. K. Maramorosch and K. F. Harris), pp. 181–200. Academic Press, Inc., New York.

IRWIN, M. E., and RUESINK, W. G. 1986, Vector intensity: a product of propensity and activity. *In* Plant Virus Epidemics: Monitoring, Modelling and Predicting Outbreaks (ed. G. D. McLean, R. G. Garrett, and W. G. Ruesink), pp. 13–33. Academic Press, Sydney.

IRWIN, M. E., and SCHULTZ, G. A. 1981. Soybean mosaic virus. FAO Plant Prot. Bull. 29:41-55.

JAYASENA, K. W., and RANDLES, J. W. 1984. Patterns of spread of the non-persistently transmitted bean yellow mosaic virus and the persistently transmitted subterranean clover red leaf virus in *Vicia faba*. Ann. Appl. Biol. 104:249-260.

JENKINSON, J. G. 1955. The incidence and control of cauliflower mosaic in broccoli in south-west England. Ann. Appl. Biol. 43:409-422.

JUTSUM, A. R., COLLINS, M. D., PERRIN, R. M., EVANS, D. D., DAVIES, R. A. H., and RUSCOE, C. N. E. 1984. PP321—a novel pyrethroid insecticide. Proc. 1984 Br. Crop Prot. Conf. Pests Dis. 1:421-428.

KENNEDY, J. S., BOOTH, C. O., and KERSHAW, W. J. S. 1961. Host finding by aphids in the field. III. Visual attraction. Ann. Appl. Biol. 49:1-21.

KOUSALYA, G., AYYAVOO, R., KRISHNAMURTHY, C. S., KANDASWAMY, T. K., and BHASKARAN, S. 1971. Effect of spacing roguing and weeding on the incidence of rosette disease of groundnut with observations on the aphid vector *Aphis craccivora*. Madras Agric. J. 58:495-505.

KRANZ, J. 1974. The role and scope of mathematical analysis and modeling in epidemiology. *In* Epidemics of Plant Diseases, Mathematical Analysis and Modeling (ed. J. Kranz), pp. 7-54. Springer-Verlag New York, Inc., New York.

KRING, J. B. 1967. Alighting of aphids on colored cards in a flight chamber. J. Econ. Entomol. 60:1207-1210.

LEWIS, T. 1965a. The effects of shelter on the distribution of insect pests. Sci. Hortic. 17:74-84.

LEWIS, T. 1965b. The effects of an artificial windbreak on the aerial distribution of flying insects. Ann. Appl. Biol. 55:503-512.

LEWIS, T. 1965c. The effect of an artificial windbreak on the distribution of aphids in a lettuce crop. Ann. Appl. Biol. 55:513-518.

LEWIS, T. 1966a. An analysis of components of wind affecting the accumulation of flying insects near artificial windbreaks. Ann. Appl. Biol. 58:365-370.

LEWIS, T. 1966b. Artificial windbreaks and the distribution of turnip mild yellows virus and *Scaptomyza apicalis* (Diptera) in a turnip crop. Ann. Appl. Biol. 58:370-376.

LEWIS, T. 1967. The horizontal and vertical distribution of flying insects near artificial windbreaks. Ann. Appl. Biol. 60:23-31.

LEWIS, T. 1969. Factors affecting primary patterns of infestation. Ann. Appl. Biol. 63:315-317.

LIQUIDO, N. J., and IRWIN, M. E. 1986. Longevity, fecundity, change in degree of gravidity and lipid content with adult age, and lipid utilisation during tethered flight of alates of the corn leaf aphid. Ann. Appl. Biol. 108:449-459.

MATTHEWS, R. E. F. 1981. Plant Virology, 2nd ed. Academic Press, Inc., New York.

MOERICKE, V. 1952. Farben als Landereise für geflügelte Blattläuse (Aphidoidea). Z. Naturforsch. 7:304-309.

MOERICKE, V. 1955. Uber die Lebensgewohnheiten der deflügeten Blattäuse (Aphidina) unter besonderer Berucksichtigung des Verhaltens beim Landen. Z. Angew. Entomol. 27:29-91.

MOERICKE, V. 1962. Uber die optische Orientierung von Blattläusen. Z. Angew. Entomol. 50:70–74.

MÜNSTER, J., and MURBACH, R. 1952. L'application d'insecticides contre les pucerons vecteurs des viroses de la pomme de terre peut-elle garantir la production de plantes de qualité? Rev. Romande Agric. Vitic. Arboric. 8:41–43.

PIRONE, T. P. 1981. Efficiency and selectivity of the helper-component-mediated aphid transmission of purified potyviruses. Phytopathology 71:922–924.

PIRONE, T. P., and MEGAHED, E. S. 1966. Aphid transmissibility of some purified viruses and viral RNAs. Virology 30:631–637.

QUIOT, J. B., LABONNE, G., and MARROU, J. 1982. Controlling seed and insect-borne viruses. *In* Pathogens, Vectors and Plant Diseases: Approaches to Control (ed. K. F. Harris and K. Maramorosch), pp. 95–122. Academic Press, Inc., New York.

RUESINK, W. G., and IRWIN, M. E. 1986. Soybean mosaic virus epidemiology: a model and some implications. *In* Plant Virus Epidemics: Monitoring, Modelling and Predicting Outbreaks (ed. G. D. McLean, R. G. Garrett, and W. G. Ruesink), pp. 295–313. Academic Press, Sydney.

SCHULTZ, G. A., IRWIN, M. E., and GOODMAN, R. M. 1983. Factors affecting aphid acquisition and transmission of soybean mosaic virus. Ann. Appl. Biol. 103:87–96.

SCHULTZ, G. A., IRWIN, M. E., and GOODMAN, R. M. 1985. Relationships of aphid landing rates to the field spread of soybean mosaic virus. J. Econ. Entomol. 78:143–147.

SCOTT, R. W., and ACHTEMEIER, G. L. 1987. Estimating pathways of migrating insects carried in atmospheric winds. Environ. Entomol. 16:1244–1254.

SHANKS, C. H., and CHAPMAN, R. K. 1965. The effects of insecticides on the behavior of *Myzus persicae* Sulzer and its transmission of potato virus Y. J. Econ. Entomol. 58:79–83.

SIMONS, J. N. 1957. Effects of insecticides and physical barriers on field spread of pepper vein-banding mosaic virus. Phytopathology 47:139–145.

SIMONS, J. N. 1958. The effect of movements of winged aphids on transmission of a non-persistent aphid-borne virus. Proc. 10th Int. Congr. Entomol. Montreal, 1956, pp. 229–231.

SIMONS, J. N. 1960. Factors affecting field spread of potato virus Y in south Florida. Phytopathology 50:424–428.

SLYKHUIS, J. T., ZILLINSKY, F. J., YOUNG, M., and RICHARDS, W. R. 1960. Notes on the epidemiology of barley yellow dwarf virus in Eastern Ontario in 1959. Plant Dis. Rep. Suppl. 262:317–322.

STOREY, H. H., and RYLAND, A. K. 1957. Viruses causing rosette and other diseases in groundnuts. Ann. Appl. Biol. 45:318–326.

SWENSON, K. G. 1968. Role of aphids in the ecology of plant viruses. Annu. Rev. Phytopathol. 6:351–374.

TAYLOR, L. R. 1951. An improved suction trap for insects. Ann. Appl. Biol. 38:382–591.

THRESH, J. M. 1974. Temporal patterns of virus spread. Annu. Rev. Phytopathol. 12:111–120.

THRESH, J. M. 1976. Gradients of plant virus diseases. Ann. Appl. Biol. 82:381–406.

THRESH, J. M. 1980. An ecological approach to the epidemiology of plant virus diseases. *In* Comparative Epidemiology (ed. J. Palti and J. Kranz), pp. 57–70. Pudoc, Wageningen, The Netherlands.

WALLIN, J. R., and LOONAN, D. V. 1971. Low level jet winds, aphid vectors, local weather and barley yellow dwarf virus outbreaks. Phytopathology 61:1068–1070.

ZEYEN, R., STROMBERG, E., and KUEHNAST, E. 1978. Research links MDMV epidemics to aphid flights. Minn. Sci. 33:10–11.

3

Mechanisms of Arthropod Transmission of Plant Viruses

Implications for the Spread of Disease

P. H. Berger* and R. S. Ferriss

INTRODUCTION

The means by which a pathogen spreads from host to host influences the spatial and temporal dynamics of resultant disease. For practically all vector-borne plant viruses, movement and transmission are dependent on association with an appropriate vector. In this chapter we discuss the mechanisms of virus transmission by arthropod vectors and how they affect spatial patterns and dynamics of plant virus diseases. We concentrate on arthropod-borne viruses (phytoarboviruses), since the great majority of plant viruses have arthropod vectors. Diseases caused by phytoarboviruses are of significant economic importance, and their epidemiology has been studied most intensively.

Research on phytoarbovirus epidemiology has been conducted from a number of different perspectives. Arthropod behavior and physiology, processes involved in virus–vector and virus–plant associations, biometeorology, practical control measures, and general plant disease epidemiology have all provided starting points for investigations seeking to increase understanding of the dynamics of these diseases in populations of plants. In the last 20 years a number of workers have integrated information on the dynamics and spatial

Note: Department of Plant Pathology, University of Kentucky, Lexington, KY 40546, US. *Current address: Dept. of Plant, Soil, and Entomological Sciences, University of Idaho, Moscow, ID 83843, US.

distribution of plant diseases (Campbell and Noe, 1985; Fleming et al., 1982; Gregory, 1968; Jeger, 1983; Kampmeijer and Zadoks, 1977; MacKenzie et al., 1985; McCartney and Fitt, 1985; Thresh, 1976; Wallace, 1978; Chapter 5, this volume), behavior and movement of arthropods (Bryant, 1969; Carter et al., 1980; Cook, 1967; Delong, 1971; Eastop, 1977; Gilbert, 1982; Gutierrez et al., 1974, 1984; Johnson, 1967, 1969; Madden, 1985; Pedgley, 1982; Raworth, 1984; Shiyomi, 1974; Taylor, 1977, 1984, 1985; Taylor and Taylor, 1977), epidemiology of plant virus diseases (Bos, 1981; Chiykowski, 1981; Duffus, 1977; Harris, 1978; Harrison, 1977, 1981; Irwin and Goodman, 1981; Madden and Campbell, 1986; McLean et al., 1986; Plumb, 1983; Raccah, 1986; Rose, 1978; Swenson, 1968; Thresh, 1974a,b, 1976, 1981, 1982, 1983a,b; Watson, 1967; Zitter, 1977), and plant virus transmission mechanisms (Ananthakrishnan, 1980; Bird and Maramorosch, 1978; Conti, 1985; Costa, 1969; Everett and Lamey, 1969; Fukushi, 1969; Fulton et al., 1980; Garrett, 1973; Granados, 1969; Harris, 1979, 1981, 1983; Ling, 1969; Milne and Lovisolo, 1977; Muniyappa, 1980; Nielson, 1968; Oldfield, 1970; Paliwal, 1980a; Peters, 1973; Pirone, 1969; Pirone and Harris, 1977; Rochow, 1969b; Rovainen, 1973, 1980; Sinha, 1973; Slykhuis, 1969, 1980; Sylvester, 1980; Walters, 1969; Watson and Plumb, 1972).

In this chapter we attempt to synthesize this large body of information with respect to a further perspective: how events in vectors at the tissue (and sometimes molecular) level lead to observed events, particularly spatial, which can be observed in plant populations. Because movement of viruliferous vectors is necessary for phytoarbovirus movement, this perspective addresses a primary chain of causality between different levels of biological organization. Since many processes affect this chain of causality, making it the focus of attention provides a framework for organizing information.

In the following sections we review the mechanisms of phytoarbovirus transmission and the processes involved in phytoarbovirus epidemics, and discuss possible effects of transmission mechanisms on disease spread and control. The huge volume of potentially relevant literature makes it impossible to cite all publications relating to the topic of this chapter. We have, however, attempted to provide representative citations where appropriate.

ELEMENTS OF PLANT VIRUS DISEASE EPIDEMICS

The dynamics of virus disease in a population of plants results from numerous interactions between vectors, plants, and virions. To obtain some understanding of this complex process, it is necessary to break the process into simpler, more comprehensible elements. The elements that have received the most attention are those that can be measured to characterize temporal associations of viruses with vectors and plants.

For transmission to occur, a vector must acquire virus from an infected plant, retain it for a sufficient period to become infective, and transmit it to an uninfected plant host. These fundamental steps in the transmission process are the bases of the temporal variables by which a virus–vector relationship can be characterized: acquisition, latent, retention, and inoculation periods (Table 1). The acquisition and inoculation periods can be defined by either the time that a vector resides on a plant (residence or access time) or the time that a vector feeds or probes. The latent period is the total time between vector acquisition of a virus and initial transmission and is also sometimes referred to as the incubation period. The retention time includes latency and is the total time that a vector remains infective. Some viruses have no latent period in their vectors, and the duration of the retention time can be highly dependent on whether or not a vector is actively feeding or probing. All vectors containing virus are considered to be viruliferous. Vectors that can transmit virus under suitable conditions are considered to be inoculative. Thus all infectious vectors are viruliferous, but viruliferous vectors may not be infectious. The duration of each of these periods varies greatly among different viruses (Table 2), and can vary for different vector species of the same virus or for different strains of a virus. Within a population of vectors, individuals may require

TABLE 1 Definitions of Terms Used in Describing Virus–Vector Characteristics

Acquisition time. The time required for a vector to acquire virus from an infectious plant. Acquisition time is usually reported in terms of an acquisition threshold (the minimum time required by any vector to acquire virus) time using either actual feeding or probing times or access times.

Inoculation time. The minimal feeding time required to effect transmission. Like acquisition time, inoculation time is reported in terms of either feeding or access.

Inoculative. Meaning that a vector is capable of inoculating a suscept with virus; infectious.

Latent period. The minimal period of time, following acquisition, during which a vector is not inoculative (i.e., a preinfective period).

Noncirculative. A term usually applied to non- or semipersistent transmission, meaning that virus does not circulate in the body of the insect.

Probe. Insect test feeding, usually to determine the suitability of potential host. Frequently of short duration.

Residence time. Duration of the period for which a vector resides (or is allowed to reside) on a plant.

Retention time. The maximal time that a vector retains inoculativity; persistence.

Transmission efficiency. Proportion of a vector population that transmits virus. Represents the ability of a given population or sample of vectors to transmit a particular virus; usually expressed as a percentage.

Transovarial. Transmission of virus to young via eggs.

Transstadial. Through a molt; a virus is transstadially transmitted in a vector that retains inoculativity following a molt.

Viruliferous. A virus-carrying (or producing) vector; not necessarily inoculative.

different periods of time to complete each step in the transmission process (Figs. 1–3). Consequently, the duration of each period (latency, retention, and transmission) can be expressed as either the minimum time needed for at least one vector to complete the step (the threshold time) or the time needed for a proportion of the population to complete the step (e.g., the time at which 50% of a population completes the latent period). Measured threshold times are highly dependent on the size of the sample taken from the vector population; for example, a shorter threshold will usually be recorded for a sample of 100 vectors than for one of 10 vectors. However, the concept has been used widely, and threshold times are often very different for different virus groups. Consequently, published threshold times may be of value in measuring transmission characteristics even though they are of questionable validity. The proportion of a vector population able to transmit a given virus defines its transmission (or vector) efficiency. Transmission efficiency under a particular set of conditions (such as residence or feeding time) can be used to define the efficiency of acquisition or transmission.

In addition to time periods characteristic of virus–vector relationships, the epidemiology of a plant virus disease is also dependent on the virus–plant relationship. Although knowledge of the time from inoculation to symptom expression can be useful in estimating when infection occurred in field situations, time from vector transmission to when the plant can serve as a virus source has a more direct effect on the course of an epidemic. Once the latent period is completed, the plant continues to be a source of virus throughout its life. However, the ease with which a vector can acquire virus can decline with plant age, due to a decrease in virus availability, titer, or decreased palatability to the vector.

Vector behavior influences the rate and extent of virus spread. For some vectors, periodic migrations are integral parts of their life cycle, and these vectors may be introduced into crop regions at approximately the same time each year (Annand et al., 1932; Chiykowski and Chapman, 1965; Gutierrez et al., 1971; Johnson, 1969; Pedgely, 1982; Way et al., 1981). For other vectors, long-range spread is rare or associated with intermittently occurring weather phenomena (Close and Tomlinson, 1975; Kisimoto, 1976; Rosenberg and Magor, 1983; Wiktelius, 1977). Within a field, vector activity can range from nearly sessile to highly mobile, depending on species, morphological forms of the vector present (winged versus nonwinged aphids, and short-winged versus long-winged leafhoppers), local weather conditions, host conditions, and other factors (Broadbent, 1953; Gilbert, 1982; Harrewijn et al., 1981; Kennedy et al., 1959; Laird and Dickson, 1963; Shanks, 1965).

The spatial pattern of diseased plants can be characterized in a number of ways (Campbell and Noe, 1985; Gray et al., 1986; Gregory, 1968; Laird and Dickson, 1963; Madden and Campbell, 1986; Madden et al., 1982; McCartney and Fitt, 1985; Nicot et al., 1984; Taylor, 1984; Thresh, 1976; Chapter 5, this volume). Maps showing locations of virus-diseased and healthy plants

TABLE 2 General Characteristics of Phytoarbovirus Transmission

Virus–Vector Relationship	Vector Type[a]	Representative Virus Groups/Viruses[b]	Virus Particle[c]	"Typical" Transmission Characteristics					
				Acquisition Threshold	Inoculation Threshold	Retention Time	Latent Period	Transstadial Passage	Transovarial Passage
Nonpersistent	Ap	Potyviruses	FR	5 sec	5 sec	6 h +	0	–	–
	Ap	Carlaviruses	FR	5 sec	5 sec	6 h +	0	–	–
	Ap	Caulimoviruses[d]	S	5 sec	5 sec	6 h +	0	–	–
	Ap	AMV	B	5 sec	5 sec	6 h +	0	–	–
Semipersistent	Ap	Closteroviruses	FR	5–15 min	5–15 min	3 d	0	–	–
	Lh	MCDV; RTV	S	15–30 min	15 min–2 h	4–6 d	0	–	–
	Mb	CSSV	B	1–4 h	<15 min	4 d	0?[e]	+?	–
Circulative	Wf	Geminiviruses	S	5–30 min	10–30 min	2 d–life	4–24 h	+	–
	Cl	Tymoviruses	S	>5 min	<1 h	2–20 d	0?	–	–
	Cl	Comoviruses	S	>5 min	<1 h	2–20 d	0?	–	–
	Cl	Sobemoviruses	S	>5 min	<1 h	2–20 d	0?	–	–
	Cl	Bromoviruses	S	>5 min	<1 h	2–20 d	0?	–	–

	Mi	Potyviruses	FR	15 min	15 min	9 d	+?	+	–
	Ap	Luteoviruses	S	>5 min	>10 min	Weeks to life	24 h	+	–
		PEMV	S	15–120 min	20–30 s		10 h	+	–
	Lh	Geminiviruses	S	1 min +	1 min +	Weeks to life	4–48 h	+	–
Propagative	Th	TomSWV	S	15 min	5 min	Weeks to life	4–10 d	+	–
	Ap/Lh	Rhabdoviruses	B	← 30 sec + →		Weeks to life	4–24 d	+	+
	Lh	MRFV, OBDV	S	15 min–6 h	8 h	Weeks to life	3–47 d	+	–
	Lh/Ph	Reoviruses[f]	S	30 min–24 h	5 min–24 h	Weeks to life	4–60 d	+	±
	Ph	RSV group	F	15 min	3 min	Weeks to life	5–21 d	+	+

[a]Key to vector types: Ap, aphid; Lh, leafhopper; Ph, planthopper; Mb, mealybug; Wf, whitefly; Cl, beetles; Mi, mites; Th, thrips.

[b]Follows classification used in Commonwealth Mycological Institute/Association of Applied Biologists, Descriptions of Plant Viruses.

[c]Key to virus particle types: S, spherical or isometric; B, bacilliform; FR, flexous rod; F, filamentous.

[d]Some members of group reported to be transmitted bimodally.

[e]A ? indicates that this property is not well characterized.

[f]Reovirus subgroup 1 is transmitted by leafhoppers and subgroups 2 and 3 by planthoppers. The frequency of transovarial transmission in planthoppers is low or none.

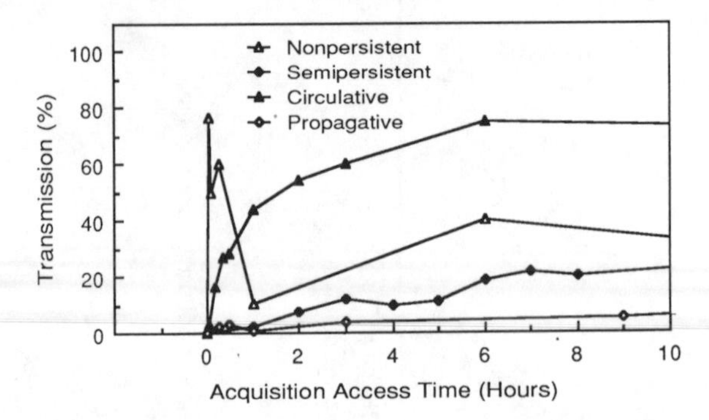

Figure 1 Effects of acquisition access time on percent transmission of phytoarboviruses with different transmission mechanisms: the nonpersistent henbane mosaic virus (Watson, 1936), the semipersistent beet yellows virus (Sylvester, 1956), the circulative sugar beet curly top virus (Bennett and Wallace, 1938), and the propagative oat blue dwarf virus (Banttari and Zeyen, 1970). Percent transmission values represent the percentage of individual vectors that successfully inoculated test plants.

can be analyzed by a number of different statistical methods (Chapter 5). Although each type of analysis has its own assumptions and procedures, most of them result in the identification of a data set as representing a pattern that is random, aggregated, or regular. Identifying a pattern as random usually exhausts available information about spatial pattern. If a pattern is aggre-

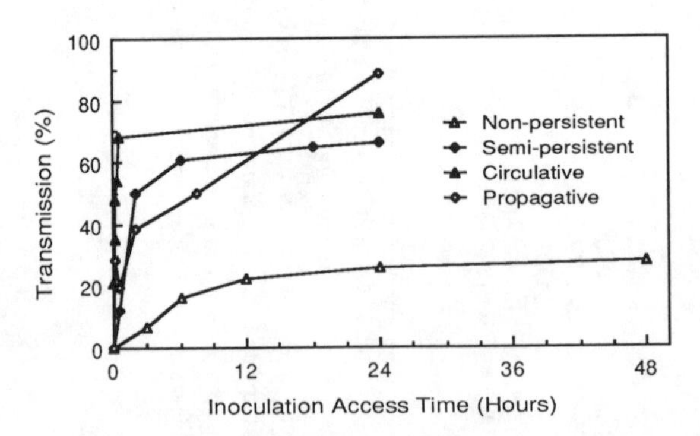

Figure 2 Effects of inoculation access time on percent transmission of phytoarboviruses with different transmission mechanisms: the nonpersistent henbane mosaic virus (Watson, 1936), the semipersistent beet yellows virus (Sylvester, 1956), the circulative sugar beet curly top virus (Bennett and Wallace, 1938), and the propagative transitory yellowing virus of rice (Chiu et al., 1968). Percent transmission values represent the percentage of individual vectors that successfully inoculated test plants.

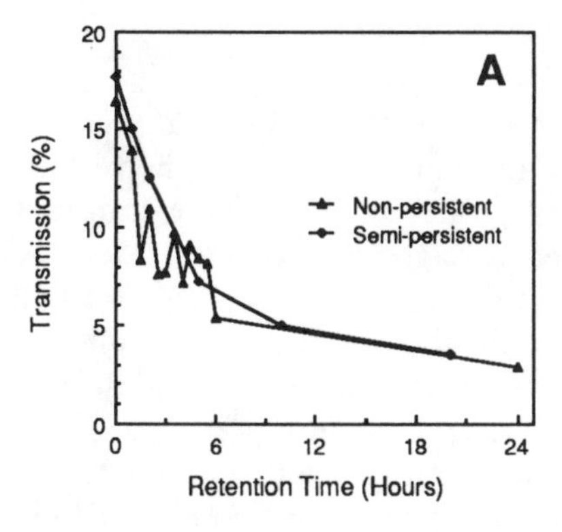

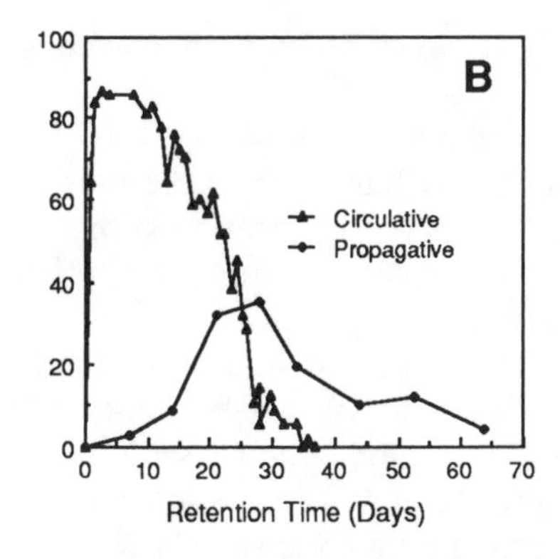

Figure 3 Effects of retention time on percent transmission of phytoarboviruses with different transmission mechanisms: (A) nonpersistent maize dwarf mosaic virus (Berger & Zeyen, 1987), and semipersistent beet yellows virus (Watson, 1946); (B) circulative pea enation mosaic virus (Sylvester, 1967), and propagative oat blue dwarf virus (Banttari, and Zeyen, 1970). Percent transmission values represent the percentage of individual vectors that successfully inoculated test plants.

gated, the particular pattern of aggregation can be examined further (Chapter 5). The most common types of aggregated patterns are those where diseased plants are clustered in foci, and those in which diseased plants are clustered near the edge of the fields. Uniform (or regular) distributions are rarely of concern in plant disease epidemiology. The effects of patterns of initial disease are discussed in the section on phytoarbovirus epidemiology.

MECHANISMS OF TRANSMISSION

The system commonly used to categorize plant virus transmission mechanisms evolved along with our understanding of virus–vector relationships. The first widely accepted system divided viruses into *persistent* and *nonpersistent* categories on the basis of the length of the retention period (Watson and Roberts, 1939). Subsequently, a *semipersistent* category was added (Sylvester, 1956), and the persistent category was recognized as consisting of two separate types of transmission: *propagative,* for viruses that multiply in the vector, and *circulative,* for viruses having a long retention time but not multiplying in vectors (Black, 1959). The term "circulative" has sometimes been used as a synonym for "persistent" (Kennedy et al., 1959). As knowledge about virus–vector relationships has increased, it has become possible to sudivide the four basic categories. However, from an epidemiological viewpoint, four categories are sufficient to provide a basic framework for discussing the implications of transmission mechanisms (Table 2).

Nonpersistent Transmission

Nonpersistent transmission is the most prevalent type of transmission in nature and has received the most attention. Only aphids serve as vectors in true nonpersistent transmission. Several groups of viruses are transmitted in this manner, including the potyviruses, carlaviruses, cucumoviruses, caulimoviruses, and alfalfa mosaic virus [the virus classification system used in this chapter is the system used by the International Committee for Taxonomy of Viruses (Harrison and Murant, 1985; Matthews, 1982; Shepherd, 1977)].

Typically, nonpersistent viruses are acquired during brief acquisition probes, lasting but a few seconds; subsequent transmission efficiency decreases as probing time increases after the first few minutes. This is thought to be a result of insects reaching host phloem tissue (where there is little or no virus to acquire) during long probes, but there may be a behavioral component as well (Nault and Bradley, 1969; Smith, 1965; Watson, 1936; 1946; Watson and Roberts, 1939, 1940). The retention periods of nonpersistent viruses are also short, usually a matter of hours; however, retention periods are more a function of sequence than of time, since virions may be lost from vectors (and inoculated into the host) during the first probe after acquisition. Consequently, retention time can be very short if the vector probes or feeds on plants immediately after acquisition, but can be relatively long (24 h or more) if the vector does not probe after acquisition (such as during long-distance movement). Transmission to multiple plants (in a series) can occur, but usually the frequency of serial transmission is low (Sylvester, 1954; Watson, 1972; Watson and Roberts, 1940). Threshold inoculation times can be quite short, and maximum transmission efficiency is often reached relatively quickly. The short times required for optimum acquisition and transmission are unique characteristics in nonpersistent transmission. There is no detectable latent period, transstadial passage, or transovarial passage (Table 1). There is no evidence for entry or circulation of virus in the hemocoel. Microinjection of virus into aphids does not result in transmission.

It was originally thought that nonpersistent transmission occurred as a result of aphid mouthpart contamination (Bradley, 1964; Bradley and Ganong, 1955a,b), and it was proposed that the term "stylet-borne" be used for this type of transmission (Kennedy et al., 1962). An alternative hypothesis is that transmission results from ingestion and egestion of plant sap containing virus (Harris and Bath, 1973; Harris, 1977). There is evidence that gross contamination of the stylets does not occur in the case of the potyviruses (Berger and Pirone, 1986; Gamez and Watson, 1964; Harris, 1977; Watson and Plumb, 1972; Watson and Roberts, 1940).

For the potyviruses and caulimoviruses, there is a helper component (HC), a viral genome-encoded protein necessary for aphid transmission (Armour et al., 1983; Govier and Kassanis, 1974; Lung and Pirone, 1973, 1974; Pirone, 1977; Pirone and Thornbury, 1984; Woolsten et al., 1983). There is evidence that HC may act as a virus–vector binding agent in portions of the stylets and food canal anterior to the esophageal valve (Berger and Pirone, 1986). In association with virus coat protein or perhaps a vector component, HC may also play a role in vector specificity. There is no evidence for the existence of an HC in either the cucumovirus, carlavirus, or the alfalfa mosaic virus groups (Pirone, 1977; Pirone and Megahed, 1966), yet their overall transmission characteristics are quite similar to groups with HC. There is some evidence for involvement of viral coat protein in transmission and vector specificity of cucumber mosaic virus (Gera et al., 1979; Mossop and Francki, 1977). Based on evidence supporting a binding mechanism hypothesis for the mode of action of the potyvirus HC (Berger and Pirone, 1986), it is tempting to assume that transmission mechanisms for nonpersistent viruses with and without helper components are similar in the manner in which virus attaches to binding sites and is eluted from these binding sites during egestion. The precise mechanism(s) controlling uptake, retention, release of virus, and vector specificity are not completely understood, but there is good evidence that HC and perhaps coat proteins play a role in some or all of these processes in the potyviruses and probably in the caulimoviruses (Pirone and Thornbury, 1983).

Many nonpersistent viruses can be transmitted by several aphid species. Frequently, there are differences in transmission efficiency when direct comparisons are made of different vectors on the same host or different hosts with the same vector (e.g., Bancroft et al., 1966; Bawden and Kassanis, 1946, 1950; Bradley and Rideout, 1953; Knoke and Louie, 1981; Messieha, 1967; Sylvester and Simons, 1951). Aphid behavior can vary from species to species, or within the same species on different hosts. Thus it is likely that differences in transmission ability among different vector species are due partially to behavioral differences. However, physiochemical differences in the ability of viruses to associate with vectors' stylets and/or foregut (via HC and/or capsid protein) may contribute to differences between vector species. Different forms and biotypes of the same vector species may also have different transmission efficiencies, possibly due to differences in biochemical properties of the vector (Berger et al., 1983; Orlob, 1962; Simons, 1959; Sohi and Swenson, 1964).

Semipersistent Transmission

Plant virus groups transmitted semipersistently are the aphid-transmitted closterovirus group (Bar-Joseph et al., 1979; Lister and Bar-Joseph, 1981), mealybug-transmitted viruses (Entwistle, 1973; Rovainen, 1973, 1980), and leafhopper-transmitted maize chlorotic dwarf virus (MCDV) groups (comprised of MCDV and rice tungro virus) (Galvez, 1971; Gingery et al., 1978, 1981). Anthriscus yellows virus (AYV) is transmitted semipersistently by aphids but has not been placed in a definite virus grouping. Caulimoviruses and pea seed-borne mosaic virus (a potyvirus) are thought to be transmitted either nonpersistently or semipersistently, depending on vector species (Chalfant and Chapman, 1962; Lim and Hagedorn, 1977). Such transmission may, however, be due to peculiarities in the feeding behavior of the vectors that transmit these viruses, rather than real differences in mechanisms of transmission (Pirone and Harris, 1977).

Most semipersistent viruses are acquired in as little as 5 to 30 min. Maximum acquisition usually occurs after acquisition times of 12 to 24 h or more. Inoculation times are longer than those for nonpersistently transmitted viruses (Sylvester, 1969), and threshold inoculation times of 15 min are common. Retention times are generally in the range of 2 to 3 days and are usually less than 6 days (Bar-Joseph and Murant, 1982; Galvez, 1971; Ling, 1966; Nault et al., 1973; Russell, 1970; Watson, 1946). The citrus tristeza closterovirus is transmitted serially, but at a low level, and presumably spreads in a manner similar to nonpersistent viruses (Raccah et al., 1976, 1978). For rice tungro virus, higher temperatures result in shorter retention times and greater transmission efficiency (Ling and Tiongco, 1979). There is no latent period for semipersistent viruses and inoculativity is lost following a molt. Transovarial passage is unknown.

The precise mechanism of semipersistent transmission is unknown, but it is apparently similar to nonpersistent transmission in terms of an ''ingestion–egestion'' mechanism (Harris, 1977, 1979). Murant (1978) and Murant et al. (1976) have shown that anthriscus yellows virus adsorbs to specific sites in the foregut of the aphid, *Cavariella aegopodii.*

Mealybugs transmit a number of economically important viruses, including grapevine leafroll closterovirus and cacao swollen-shoot virus (Brunt, 1970; Engelbrecht and Kasdorf, 1985; Owusu, 1983; Posnette, 1947; Rosciglione et al., 1983; Thresh, 1980). Mealybug-transmitted viruses behave differently from aphid- and leafhopper-borne semipersistent viruses. Mealybug transmission has characteristics of nonpersistent, semipersistent, and persistent transmission, but most closely resembles semipersistent transmission (Rovainen, 1980). A preacquisition starvation effect (usually, a characteristic of nonpersistent rather than persistent transmission) and transstadial passage (a characteristic of persistent transmission) have been reported (Rovainen, 1976, 1980). The thresholds for acquisition and inoculation are about 4 h and 15 min, respectively. In both cases, transmission efficiency increases as inoculation or acquisition time increases. Maximum retention time is about 4 days. There is no evidence of a latent period, recovery of virus from hemolymph, or acquisition by injection of virus into the hemocoel (Harris, 1981; Rovainen, 1980). Mealybugs are sedentary insects, and the spread of a disease such as cacao swollen shoot is slow (Thresh, 1958, 1974b, 1980, 1983a).

Circulative Transmission

Several important groups of plant viruses are transmitted in a circulative manner by aphids, beetles, whiteflies, mites, or leafhoppers, including the luteoviruses, geminiviruses, bromoviruses, comoviruses, sobemoviruses, and tymoviruses. Viruses in this category are characterized by acquisition and inoculation times of from as little as 15 min up to 1 h. Generally, longer residence times on infected plants result in higher levels of inoculativity. A latent period of several hours to more than 1 day is common, except for beetle-borne viruses, where current evidence indicates that the latent period is very short or absent (Fulton et al., 1980). Latent periods are considered good evidence supporting circulative (and propagative) transmission, because viruses must move or be transported from the initial site of acquisition to the salivary glands. Thus latent periods are a reflection of the time taken for circulation processes. Transstadial passage is known for all except the beetle-borne viruses. Retention times are measured in terms of days or weeks, and inoculativity may be retained for the life of the vector. Generally, transmission efficiency decreases gradually throughout the retention period if the vector does not reacquire virus.

Whitefly transmission. Whiteflies transmit some geminiviruses, viruses in the cowpea mild mottle cluster, and sweet potato mild mottle virus. The latter two groups have properties similar to those of the carlavirus and potyvirus groups, respectively. However, the virus–vector relationships are not well established at this time (Murant et al., 1987). These viruses are receiving increasing attention from researchers, due to increasing amounts of economic loss as well as the interesting biochemical nature of the geminiviruses (Bird and Maramorosch, 1978; Costa, 1975; Goodman, 1977; Goodman et al., 1977; Mound, 1973; Muniyappa, 1980; Shivanathan, 1983). *Bemisia tabaci* is the vector of almost all known whitefly-transmitted viruses. Acquisition and inoculation thresholds range from 10 min to 8 h, depending on the virus. Latent periods vary from 4 to 24 h and retention of inoculativity can be from 2 days to life (Bird and Maramorosch, 1978). There is no evidence for transovarial passage, and although transstadial passage appears to occur, the adults are most important as vectors. Little is known about the long-distance movement of whiteflies (Johnson, 1969; Pedgley, 1982).

Beetle transmission. Four groups of plant viruses have beetle vectors: the bromoviruses, comoviruses, sobemoviruses, and tymoviruses. Complete discussions of beetle transmission are provided by Fulton et al. (1975, 1980), Selman (1973), and Walters (1969). Beetles are chewing insects with feeding behavior that is fundamentally different from that of sap-sucking insects. Generally, acquisition increases with increased feeding time, and 5-min thresholds have been reported. Inoculation times have not usually been determined; beetles can apparently transmit viruses as soon as feeding begins and there is no evidence of a latent period. Retention is variable, depending on beetle species, virus, host, environmental conditions, and amounts of tissue consumed during acquisition feeding. Retention times ranging from 2 to 20 days have been reported (Dale, 1953; Freitag, 1941a,b, 1956; Smith, 1965; Walters, 1969). Walters et al. (1972) reported that bean pod mottle virus overwintered in hibernating beetles. Virions injected into the hemocoel can render beetles inoculative.

The precise mechanism of beetle transmission is unclear at this time. The high degree of specificity between virus and vector would seem to rule out simple mouthpart contamination. Inoculativity following injection of viruses into the hemocoel and the apparent rapid movement of viruses from the midgut into the hemocoel (as well as back into the midgut) suggest a specific recognition/transport system. Additionally, it has been observed that virus occurs in beetle regurgitant at a higher concentration than in the source plant, suggesting a specific, selective process (Scott and Fulton, 1978).

Recently, it was demonstrated that ribonuclease in beetle regurgitant has an effect in determining whether some viruses will be beetle transmissible (Gergerich et al., 1983; Monis et al., 1986). The beetle-borne viruses tested were resistant to beetle-regurgitant RNase, whereas other, nontransmissible viruses were not. Although this finding suggests specificity, it is not known what the relationship is between virus in the hemocoel and inoculativity.

Mite transmission. Eriophyid mites are known to transmit only a few viruses, such as wheat streak mosaic (WSMV), agropyron mosaic, and ryegrass mosaic viruses (Oldfield, 1970; Slykhuis, 1980). These viruses have been placed in a subgrouping of the potyvirus group, but their virus–vector relationship is atypical of potyviruses.

All instars of *Eriophyes tulipae,* the vector of WSMV, are capable of transmitting virus, but adults can do so only if they acquired virus as nymphs. Transovarial passage has not been reported. Like a typical circulative virus, acquisition occurs only after feeding times exceeding 15 min, with transmission efficiency increasing as acquisition feeding time increases. Retention of inoculativity for up to 9 days has been reported (Orlob, 1966; Slykhuis, 1955).

Although the precise mechanism of transmission remains undetermined, WSMV has been found in different vector tissues (Harris, 1981). Paliwal and Slykhuis (1967) were able to localize concentrations of WSMV in portions of the *E. tulipae* midgut, while Takahashi and Orlob (1969) confirmed this observation and found virus in parenchymatous tissue around the intestine. Paliwal (1980a) reported finding virus particles in salivary glands of the mite vector. Thus it appears that WSMV is circulative and that a mechanism exists for transport from the alimentary canal into the salivaries (Paliwal, 1980a,b).

Eriophyid mites are among the smallest of arthropod vectors: adults are about 250 μm in length. They can crawl short distances, but dispersal is due primarily to factors such as wind, insects, or perhaps birds (Gibson and Painter, 1957). Wind is apparently the primary means of dispersal (Nault and Styer, 1969; Pady, 1955). There are examples of graminaceous hosts providing reservoirs of *E. tulipae* and WSMV, but the widespread cultivation of wheat is sufficient to explain the abundance of virus and vector (Conin, 1956; Gates, 1970).

Aphid and leafhopper transmission. A number of important viruses are transmitted by aphids or leafhoppers in a circulative manner. They include the aphid-borne barley yellow dwarf virus (BYDV) and potato leafroll virus (PLRV) of the luteovirus group, pea enation mosaic virus (PEMV is the only known member of its group), and the leafhopper-borne geminiviruses, particularly beet curly top virus (BCTV) and maize streak virus (MSV).

Minimum acquisition and inoculation feeding times range from 1 to 15 min and from 20 s to 15 min, respectively. Vectors will retain transmission ability through molting, but transovarial transmission is lacking. Latent periods of approximately 24 h are common, but can range from 4 to 48 h. Retention times are variable, but retention of inoculativity over a period of weeks is common. Vectors that are hemocoel microinjected with purified virus are inoculative. Latent periods and retention times are correlated with the dose of inoculum in microinjected insects (Paliwal and Sinha, 1970).

BYDV has been investigated more than any other phytoarbovirus, and what is known for BYDV (and other luteoviruses) may prove true for other aphid- or leafhopper-borne circulative viruses. There are a number of vector-specific strains of BYDV (Rochow, 1969a). Virus can be detected in aphids that do not transmit specific strains of BYDV (Rochow and Pang, 1961), and vector specificity is retained when virus is microinjected into the hemocoel (Rochow, 1969a). There is no change in vector specificity when the gut wall is punctured (Rochow and Pang, 1961). There is no evidence implicating hemolymph factors in vector specificity; thus it seems that selective uptake of virus by salivary glands is likely. Using electron microscopy, Harris et al. (1975) observed the specific association of PEMV with the salivary glands, and some of the luteoviruses have also been found to be transported into the salivary glands of vector aphids (Gildow, 1982; Gildow and Rochow, 1980). Gildow (1982) has suggested that "a cellular mechanism, involving coated-vesicle transport of virions" is responsible for movement of virus through accessory salivary gland cytoplasm, and observed a similar association between a strain of BYDV and cell membranes of the aphid hindgut (Gildow, 1985). It appears that specificity depends on both whether viruses can pass through the gut wall to the hemocoel and whether they can pass from the hemocoel into the salivary glands (Gildow, 1985; Tamada and Harrison, 1981).

Propagative Transmission

The basic difference between the circulative and propagative transmission systems is that in propagative transmission the vector is a multiplication host for the virus. With the propagative viruses the degree of vector specificity is very high. Most commonly, a single virus is transmitted by a single vector species; but many propagative viruses have wide plant host ranges. Various kinds of evidence have been used to demonstrate virus multiplication in vectors: serial injection of previously virus-free insects, so that a dilution factor is obtained that far exceeds the dilution endpoint of the initial inoculum; electron microscopy of insect tissues; demonstration of transovarial passage; propagation of virus in insect tissue culture; and circumstantial transmission evidence, such as the existence of long latent periods and a gradual increase in transmission efficiency for several weeks following the long latent period. Several reviews on propagative transmission have been published (Bennett, 1967; Delong, 1971; Forbes and MacCarthy, 1969; Francki et al., 1981; Harpaz, 1972; Harris, 1981; Kisimoto, 1973; Maramorosch and Harris, 1979; Milne and Lovisolo, 1977; Nault and Rodriguez, 1985; Nielson, 1968, 1979; Shikata, 1981; Sinha, 1973; Storey, 1939).

A number of plant viruses are propagative in their vectors, including plant rhabdoviruses, the maize rayado fino virus (MRFV) group, plant reoviruses, and the rice stripe virus group. Rhabdoviruses are transmitted either by aphids, leafhopper, or planthoppers. The MRFV group is comprised of two members, MRFV and oat blue dwarf virus (OBDV), and both are leafhopper-borne. Reoviruses are transmitted by leafhoppers and planthoppers, while viruses in the rice stripe virus (RSV) group are transmitted by planthoppers. It is possible, but not yet confirmed, that tomato spotted wilt virus (TSWV) is propagative in its thrip vectors (Francki and Hatta, 1981; Milne and Francki, 1984; Paliwal, 1979). TSWV is acquired only by larvae but is transmitted only by adults (Ie, 1970), and is not transmitted transovarially (Sakimura, 1963).

Minimum acquisition feeding times are approximately 15 min for TSWV (Ie, 1970) and OBDV (Banttari and Zeyen, 1970, 1973), 30 min to 24 h for reoviruses (Boccardo and Milne, 1984), 6 h for MRFV (Gamez, 1980), and 30 s for rhabdoviruses. Inoculation feeding times have similar ranges, although thresholds of as little as 30 s have been reported for some rhabdoviruses. Transmission efficiency is positively correlated with both acquisition and inoculation feeding times, and optimum transmission efficiency is achieved only after a relatively long period of acquisition and inoculation. The duration of the latent period is highly variable and depends on the particular virus–vector system. Latent periods from 3 to 60 days have been reported. High temperature decreases the duration of latent periods for rhabdoviruses (Peters, 1981) and may have similar effects on other viruses. Retention of inoculativity can be for the life of the insect, and virus is retained through molts. Whereas TSWV and the MRFV group are not known to be transmitted transovarially, most or all of the rhabdoviruses are thought to be so transmitted. Virus will pass to a high proportion of eggs in the reovirus subgroup 1 (leafhopper transmitted) and to none or a low percentage in subgroups 2 and 3 (planthopper transmitted) (Boccardo and Milne, 1984). Microinjection with virus or extracts of viruliferous vectors results in some of the insects becoming inoculative.

Vector populations can vary considerably in their ability to transmit a particular propagative virus (Storey, 1939; Timian and Alm, 1973). For example, Kisimoto (1967) was able to select planthopper genotypes that had high or low RSV transmission efficiency and found a positive correlation between transmission efficiency and transovarial transmission. The significance of transovarial transmission in virus spread is not entirely clear, as little is known about the effects of infection on the insect host in terms of its survival, fertility, behavior, and maintenance of virus (Fine and Sylvester, 1976). Generally, the geographic range of viruses and vectors overlap, and in most cases, the plant host range of an individual virus is similar to the host feeding range of its vector species. Although it would seem logical to conclude that propagative viruses have coevolved with both their plant and insect hosts, the wide plant host ranges but narrow insect host ranges of these viruses indicate that these are probably insect viruses that evolved an ability to infect plants.

Based on the general transmission characteristics, and the fact that virus multiplies in vector tissues, including the salivary glands, it appears that inoculation results from egestion of virus with saliva. The levels of physiological specificity postulated for circulative viruses also appear to apply to propagative viruses, with the addition of another level of specificity defined by virus multiplication in the vector.

PHYTOARBOVIRUS EPIDEMIOLOGY

The spread of a plant virus disease can be subdivided into two major categories: within-field spread and between-field (or into-field) spread. From a pragmatic standpoint, a field is a particular area where disease is monitored intensively. In this chapter we use a more conceptual definition: A field is a relatively homogeneous population of host plants within which a set of processes occurs that is different from the set associated with between-field phenomena. From this standpoint, some of what can be said in reference to a field can also be said of larger areas such as counties and crop regions. For a phytoarbovirus, between-field spread can involve quite different considerations than those of within-field spread. Between-field spread is primarily a function of virus retention time, weather, and vector behavior associated with long- and intermediate-range movement (including takeoff and landing). Within-field spread is a function of all the characteristics of transmission, as well as practically all aspects of vector reproduction, feeding, and movement.

The effects of transmission characteristics on virus spread can be envisaged in a number of ways. If one considers the events that can occur after a single vector acquires virus, it is possible to develop simple, informal models of how spread might be affected. Such an approach is well suited to most factors involved with between-field spread and some of those factors involved with within-field spread. However, more formal models are necessary to help evaluate the consequences of many vectors acquiring and inoculating virus in a population of plants. In the following sections, we discuss the general characteristics of phytoarbovirus epidemics; deal with within-field spread in terms of single vectors and in terms of a computer model of a population of vectors in a field; and consider the separate but related process of between-field spread.

Patterns of Initial Disease

Inoculum initiating a virus disease epidemic in a field can come from a number of sources. Plant propagative materials (such as seeds, transplants, etc.) can already be infected at planting; virus can be present within the field in soilborne vectors, weeds, or volunteer crop plants; or vectors can introduce virus from sources outside the field (Thresh, 1980, 1982). If propagative materials from a single source (e.g., seedlot) are planted, initially infected plants will usually be randomly distributed throughout the field. If the virus is present in the field as sources such as soilborne vectors or weeds, the initial spatial pattern of disease may reflect the spatial pattern of the sources (see Chapter 5). Soilborne fungi, nematodes, and weeds usually have aggregated spatial patterns (Campbell and Noe, 1985).

The initial disease pattern that results when infectious vectors introduce virus from outside a field depends on the pattern in which the vectors land and their behavior after landing. If the source is near a field border (such as in adjacent fields, ditches, or fence rows), plants near that border will be more likely to be landed on by infectious vectors. This can result in an initial spatial pattern that is aggregated around the virus source. If the source is relatively far away from the field, gradients across the field due to differences in source proximity may be undetectable; the pattern in which vectors settle will depend primarily on the environment and vector behavior, and may appear to be random. Plant color and density can have a significant effect on vector landing behavior and may induce vectors to concentrate on certain plants or areas of the field (Cartier, 1963; Irwin and Ruesink, 1986; Chapter 2). If vectors do not discriminate between plants, the pattern of vector deposition will be affected primarily by weather variables such as wind (Takami and Smith, 1972). The influence of wind on deposition can be either neutral (resulting in a random pattern) or can result in a denser deposition of vectors in some areas, such as a particular distance behind a windbreak (Quiot et al., 1979).

The timing and extent of inflights of infectious vectors can vary greatly between epidemics (Taylor, 1986). For some diseases there may be only a single major influx of vectors. The subsequent course of the epidemic then depends entirely on within-field movements of vectors already present. For other epidemics, inoculative immigrant vectors may be introduced continuously throughout the entire growing season.

If vectors are active in a field, plants that were infected by initial inoculum can become inoculum sources and within-field spread can occur. The period from when a plant becomes infected until inoculum produced in it is transmitted to other plants can be considered to constitute a generation or infection cycle. If vectors acquire and inoculate virus as quickly as possible, the duration of a generation will be the sum of the plant and vector latent periods. If many infection cycles occur sequentially within a field during a growing season, the resultant epidemic can be considered to be at least partially polycyclic (Vanderplank, 1963; Zadoks and Schein, 1979). A completely polycyclic epidemic can result if only within-field spread occurs, and a completely monocyclic epidemic can result if a steady influx of viruliferous vectors from outside the field is the only source of inoculum. Because epidemics of phytoarbovirus diseases can involve both within-field and between-field spread, a particular epidemic can be monocyclic, polycyclic, or both. Actual epidemics are subject to variations in their rate of progress due to day-to-day variations in influxes of inoculum, disease development in host plants, and vector behavior (see Chapter 2). However, no matter how uneven the progress of an epidemic, it can still be considered to be composed of monocyclic and/or polycyclic subepidemics.

Within-Field Spread

Individual vectors. The lengths of periods involved in the transmission process can have obvious effects on the patterns of plants inoculated by a single viruliferous vector. The longer the latent period, for example, the more likely it will be that a vector may travel some distance from where it acquired virus before becoming inoculative. For a nonpersistent virus, the lack of a latent period and the short required inoculation period make it likely that a vector will inoculate plants near the one from which it acquired virus. If vector movement is primarily to adjacent plants, most newly infected plants will be next to a previously infected plant and a steep disease gradient will result (Thresh, 1976). In contrast, the relatively long latent periods of propagative viruses can result in a vector moving considerable distances before it becomes inoculative. Consequently, there may be little or no development of foci around initially infected plants, and patterns of disease after within-field spread of propagative viruses would be expected to be more random than those for nonpersistent viruses.

The relative locations of plants landed on by infectious vectors can have a great effect on disease pattern: the more plants that a vector flies over without landing, the less aggregated is the resultant pattern. However, the effects of such differences will be similar for all transmission mechanisms as long as the flight time of an infectious vector is short in relation to retention time. The time that an infectious vector spends on an uninfected plant can have different implications for disease spread with different transmission mechanisms. If a vector remains inoculative for a long period, it can inoculate a number of plants after acquiring virus. If an infectious vector of a circulative or propagative virus inoculated every plant it landed on and moved primarily between adjacent plants, a tight cluster of diseased plants might be observed in the area where the vector became inoculative. However, because inoculation efficiency is usually well below 100%, a vector will not necessarily inoculate every plant that it visits. This will result in "skips" and thus a less aggregated disease pattern than that which would result if every plant were inoculated.

Effects of transmission characteristics on simulated epidemics. To examine the theoretical effects of different transmission characteristics on the spread of disease, we developed a relatively simple computer simulation model. The model is intended to simulate the events that occur after a single plant is infected by a phytoarbovirus (Table 3). It is therefore a model of the expansion of a single focus of disease. The basic time step is the length of one hypothesized residence period: the minimum time for which a vector can reside on a particular plant before moving on. A simulated epidemic begins with a single plant in a 17 plant × 25 plant field being designated as being infected. Vectors are then placed on the 425 plants at random. At each time step, each vector can make up to a specified number of moves to any adjacent plant, or it can remain stationary. Further movement is possible in the direction of the prevailing wind. After the plant latent period has elapsed, vectors on each diseased plant can acquire the virus. If a vector has been viruliferous for longer than the vector latent period but shorter than the retention period, that vector has the potential to inoculate the virus if it lands on an uninfected plant. For each vector at each time step, the values of pseudorandom variables determine the distance and direction of movement and whether virus will be acquired or inoculated. The model is based on many simplifying assumptions about virus, plant, and vector behavior. For example, plant susceptibility does not change with age; the vector population is constant; and vectors do not have any preference for infected (compared to uninfected) plants. However, the model provides a means for evaluating how disease spread is affected by particular factors, including ones that may be difficult or impossible to determine experimentally (such as the probability that an individual vector will move during a time period).

To compare the effects of transmission characteristics, a series of simulation runs was performed in which the initial conditions and parameters for each run were identical except for lengths of vector latent and infectious periods. Combinations of latent and infectious periods were used that correspond to generalized transmission characteristics for each type of transmission (Fig. 4). Although all disease patterns were significantly ($P < 0.001$) aggregated according to a runs test (Madden et al., 1982), the amount of aggregation (as measured by the number of runs of infectious or uninfected plants) differed for the different types of transmission. Aggregation was greatest for nonpersistent and semipersistent viruses, followed by circulative and then pro-

TABLE 3 Outline of Simulated Within-Field Spread[a]

I. Variables for which values are specified by the user:
  1. Duration of the latent period in the host.
  2. Duration of the latent period in the vector.
  3. Duration of the infectious period in the vector.
  4. Duration of the vector residence period.
  5. Efficiencies of acquisition and inoculation.
  6. The probability that a vector will move.
  7. The maximum distance (number of plants) that a vector can move during a time step.
  8. The number of vectors per plant.

II. Initialization:
  1. A single plant in a 17 × 25 plant field is designated as being newly inoculated at time = 0.
  2. Nonviruliferous vectors are placed on the 425 plants at random. The total number of vectors is the specified number per plant × 425.

III. At each time step (duration = the specified vector residence period):
  1. For each plant:
     A. Acquisition of virus by vectors. If the plant was infectious at the start of the time step, each nonviruliferous vector on the plant acquires virus if the value of a pseudorandom number[b] is less than the specified acquisition efficiency for that vector.
     B. Inoculation of plants by vectors. If the plant was uninfected at the start of the time step, it is inoculated if the value of a pseudorandom number is less than the specified inoculation efficiency for any infectious vector on the plant.
  2. For each vector:
     A. Movement to adjacent plants. For each of the number of moves that can occur during a time step, the probability that a vector will move to each of the four adjacent plants is one-fourth of the overall probability that a vector will move.

[a]This simulation model has been implemented in Pascal on IBM PC, Apple II, and Apple Macintosh computers. Copies of the source code and the compiled program can be obtained from the second author upon request.

[b]A number between 0 and 1 is selected by a pseudorandom procedure each time a decision is to be made (such as whether a vector will acquire virus from an infectious plant).

pagative viruses (Fig. 4). The patterns that were observed in the simulations were consistent with what would be expected from consideration of the path followed by a single viruliferous vector: Longer paths (the result of longer retention times) and longer latent periods resulted in more skips and consequently less aggregation. The nonpersistent virus was highly aggregated because each vector could inoculate only the first plant it landed on after leaving the previously infected plant from which it acquired virus. For the semipersistent virus, skips could occur in association with the low amount of serial transmission that was specified in the model: A viruliferous vector could fail to inoculate the first plant it landed on after acquiring virus, but then inoculate the second plant. Compared with the nonpersistent virus, this possibility of

Nonpersistent:

```
* * * * * * * * * * * * * * * * *
* * * * * * * * * * * * * * * * *
* * * * * * * * * * * * * * * * *
* * * * * * * * * * * * * * * * *
* * * * * * * * * * * * * * * * *
* * * * * * * * * * * * * * * * *
* * * * * * * * X 1 * * * * * * *
* * * * * X X X X X * 1 * * * * *
* * * * X X X X S X X X * * * * *
* * * * * X X X X X X X * * * * *
* * * * * X * X X X X 1 * * * * *
* * * * * * * 1 * * * * * * * * *
* * * * * * * * * * * * * * * * *
* * * * * * * * * * * * * * * * *
* * * * * * * * * * * * * * * * *
* * * * * * * * * * * * * * * * *
* * * * * * * * * * * * * * * * *
```

Semipersistent:

```
* * * * * * * * * * * * * * * * *
* * * * * * * * * * * * * * * * *
* * * * * * * * * * * * * * * * *
* * * * * * * * * 1 * * * * * * *
* * * * * * * 1 * * * * * * * * *
* * * * * * * 1 X X * * * * * * *
* * * * * * 1 1 X X X * * * * * *
* * * * * * X X X X X 1 1 * * * *
* * * * * * X X S X X * * * * * *
* * * * * * X X X 1 * * * * * * *
* * * * * * 1 X X 1 * * * * * * *
* * * * * * * * * * * * * * * * *
* * * * * * * * * * * * * * * * *
* * * * * * * * * * * * * * * * *
* * * * * * * * * * * * * * * * *
* * * * * * * * * * * * * * * * *
* * * * * * * * * * * * * * * * *
```

Circulative:

```
* * * * * * * * * * * * * * * * *
* * * * * * * * * * * * * * * * *
* * * * * * * * * * * * * * * * *
* * * * * * * * * * * * * * * * *
* * * 1 * * * * * * * * * * * * *
* * * 1 * 1 1 * 1 * * * * * * * *
* * * 1 X 1 1 1 X X X * * * * * *
* * * * * X X X X X 1 * * * * * *
* * * * 1 1 X X S 1 * * * * * * *
* * * * 1 * * 1 * 1 1 * * * * * *
* * * * * * * 1 * * * * * * * * *
* * * * * * * * * * * * * * * * *
* * * * * * * * * * * * * * * * *
* * * * * * * * * * * * * * * * *
* * * * * * * * * * * * * * * * *
* * * * * * * * * * * * * * * * *
* * * * * * * * * * * * * * * * *
```

Propagative:

```
* * * * * * * * * * * * * * * * *
* * * * * * * * * * * * * * * * *
* * * * * * * * * * * * * * * * *
* * * * * * * * * * * 1 1 * * * *
* * * * * * * * * * * 1 * * * * *
* * 1 * * * * * * * * * * * * * *
* * * * * * 1 * * X X * * * * * *
* * * * * * * 1 * X X * * * * * *
* * * * 1 X X 1 S * * * * * * * *
* * * * 1 * 1 * 1 * * * * * * * *
* * * * * 1 X 1 1 * * * * * * * *
* * * * X X X X * * * * * * * * *
* * * * X X X 1 * * * * * * * * *
* * * * * * 1 * * * * * * * * * *
* * * * * * * * * * * * * * * * *
* * * * * * * * * * * * * * * * *
* * * * * * * * * * * * * * * * *
```

serial transmission increased the rate of epidemic progress but has little or no effect on spatial pattern. For circulative and propagative viruses, long retention times resulted in each infectious vector visiting many plants; thus there were many opportunities for a vector to fail to inoculate a plant by chance, and then inoculate the next plant on which it landed. The long vector latent period for propagative viruses added more skips, resulting in even less aggregation. If vectors had moved greater distances in each movement or inoculation efficiency had been much lower, a nearly random pattern would have been observed for the propagative virus, even though only within-field spread occurred.

Disease progress in simulations was similar to what might be expected when considering individual vectors: The rate of increase in the number of infectious plants was faster for transmission types with longer retention times, and disease increase was delayed by the long latent period of the propagative virus compared with the circulative virus (Fig. 5). The relatively slow progress for the nonpersistent virus in the simulations summarized in Figs. 4 and 5 is inconsistent with reports of rapid increases in disease incidence in actual epidemics (Thresh, 1976; Zink et al., 1956). However, epidemics proceeded much more rapidly in simulations where the vector population was larger, vectors could move a distance of more than one plant during each residence period, and/or the residence was shorter (data not shown). It is possible that cases where epidemics of nonpersistent diseases proceed rapidly are due to populations of vectors that make frequent and/or extensive movements among plants. Such a high degree of vector activity could result in the rapid spread of a nonpersistent virus because of the short inoculation and acquisition thresholds.

Between-Field Spread

Vectors can start long-range movements in response to developmental cues (such as sexual maturity) and/or particular climatological events (John-

Figure 4 Locations of infectious ("X") and latently infected ("l") plants in representative simulated phytoarbovirus epidemics at the time when the number of infected plants first equaled or exceeded 30. "S" indicates the location of the initially infected plant and "*" indicates the location of uninfected plants. Values used for the vector latent and infectious periods were 0 and 1 day for nonpersistent, 0 and 2 days for semipersistent, 1 and 28 days for circulative, and 8 and 48 days for propagative. For all transmission types, the plant latent period was 8 days, the vector residence period was 1 day, acquisition and inoculation efficiencies were each 0.50, the probability of movement was 0.8, there was one vector per plant, and each vector made one move per residence period. For each transmission type, the simulated epidemic for which a map is displayed is that for which the number of rows of contiguous infected plants was nearest the mean for 20 simulated epidemics. Disease progress data for the same groups of 20 simulated epidemics are presented in Fig. 5. An outline of the simulation model is presented in Table 3.

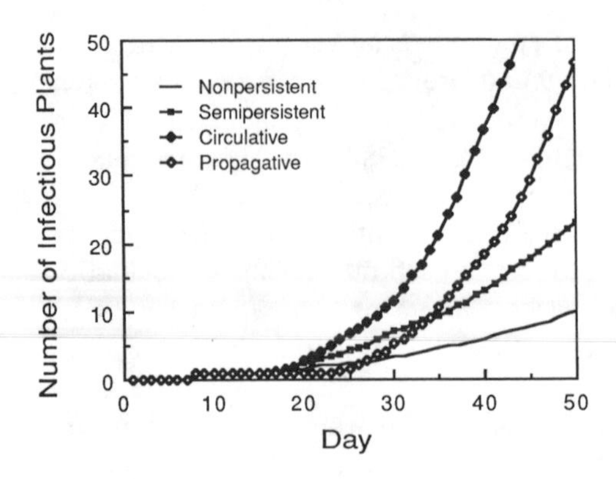

Figure 5. Disease progress (number of infectious plants) in simulated phytoarbovirus epidemics. Data are means from simulated epidemics. An outline of the simulation model is presented in Table 3.

son, 1967, 1969; Pedgely, 1982; Rose, 1978). In either case, the number of viruliferous vectors arriving at a distant area depends on both the size of the initial population and the proportion retaining virus on arrival.

Source population of vectors. A population of viruliferous vectors implies the existence of a population of infectious source plants, and thus a source epidemic. During any epidemic caused by a phytoarbovirus disease, there are changes in both the number of infected plants and the number of viruliferous vectors. If we are considering primarily within-field spread, the center of attention becomes the number and pattern of diseased plants; the number of viruliferous vectors is mainly of importance in how it affects disease in the plant population. However, if we are concerned with how suitable an epidemic is as a source of viruliferous vectors for between-field spread, attention must shift to the vector population.

It can be expected that longer retention times will result in larger proportions of vector populations being viruliferous for a given disease incidence in a plant population. If vector populations were similar on source and uninfected host plants, the short retention times of nonpersistent viruses would result in the proportion of the vector population that is viruliferous being close to the current proportion of source plants. Whether a vector carried virus as it left a field would depend on whether the last plant it probed on was a source of virus and its behavior prior to leaving that plant (Zettler and Wilkinson, 1966). For semipersistent viruses, the generally longer infectious period would be expected to result in a larger portion of the vector population being viruliferous, since a vector might have acquired virus from a number of plants, not just the last one it visited. For persistent viruses, the number of viruliferous vectors would depend on disease incidence over the course of a relatively longer period.

Retention during movement. Whether a vector will still be viruliferous after moving between fields will depend on the time taken for the journey.

The maximum duration of the retention period places an upper limit on the time for which a vector can be removed from an infectious plant and still remain viruliferous. However, the estimation of potential transit times from published retention times involves different considerations for persistent and nonpersistent viruses.

For persistent viruses, retention times probably rarely limit the distance that a vector can carry virus. Consequently, questions about how far these viruses can spread must be more concerned with how far their vectors can move. There are many cases in which the long-distance dispersal of aphids, planthoppers, and leafhoppers has been observed (Table 4), and such phenomena are well known for other insects and for some arboviruses of medical and veterinary importance (Johnson, 1969; Nuorteva and Hoogstraal, 1963; Pedgely, 1982; Sellers, 1980; Sellers et al., 1979). Although comparable movements of plant viruses have rarely been documented, the frequent long-range movement of barley yellow dwarf virus in North America provides an excellent example for a persistent (circulative) virus. There is evidence of BYDV epidemics in the United States in 1890 and 1907, a BYDV epidemic was reported in California in 1951, and in 1959 a serious epidemic of BYDV occurred in the

TABLE 4 Some Examples of Long-Distance Dispersal/Migration of Aphids, Leafhoppers, or Planthoppers

Species	Type[a]	Approximate Distance (km)	Ref.[b]
Macrosiphum miscanthi	Ap	1300	1
Aphis craccivora	Ap	320–480	4
Cinara picea	Ap	1300	2
Homoptera, Hemiptera	Ap/Lh/Ph	1100–1500	3
Sogatella furcifera	Ph	600–2000	6
Nilaparvata lugens	Ph	600–2000	5, 6
Empoasca fabae	Lh	<500	7, 8, 20
Schizaphis graminum	Ap	<800	9, 10, 22
Circulifer tenellis	Lh	<600	11, 21
Macrosteles fascifrons	Lh	1500	12–15
Cicadulina spp.	Lh	<100	16
Sogatodes spp.	Ph	200–1000	17
Myzus persicae	Ap	<150	18, 19

[a]Ap, aphid; Lh, leafhopper; Ph, planthopper.

[b]Key to references: 1, Close and Tomlinson (1975); 2, Elton (1925); 3, Holzapfel et al. (1970); 4, Johnson (1957); 5, Rosenberg and Magor (1983); 6, Kisimoto (1976); 7, Glick (1960); 8, Huff (1963); 9, Medler & Smith (1960); 10, Hodson and Cook, (1960); 11, Annand et al. (1932); 12, Chiykowski and Chapman (1965); 13, Chiykowski (1953); 14, Drake and Chapman (1965); 15, Nichiporick (1965); 16, Rose (1973); 17, Everett and Lamey (1969); 18, Wiktelius (1977); 19, Johnson (1967); 20, Pienkowski and Medler (1964); 21, Cook (1967); 22, Bruehl (1961).

north central Great Plains (Bruehl, 1961; Oswald and Houston, 1951; Thresh, 1980). It is thought that greenbugs (*Schizaphis graminum*) being dispersed from the southern United States were responsible for the 1959 epidemic (Bruehl, 1961; Hodson and Cook, 1960; Medler, 1962; Medler and Smith, 1960; Orlob and Arny, 1960). A relationship has been observed in the United States between BYDV outbreaks and low-level jet wind activity (Bruehl, 1961; Wallin and Loonan, 1971; Wallin et al., 1967). A similar situation probably exists in Europe and Australia but has yet to be confirmed (Plumb, 1983). The amount of primary infection is dependent on the number of inoculative migrant insects, duration of migration, local climatological conditions, and the condition and growth stage of the crop at the time of migration (Dean, 1974; Halbert and Pike, 1985). Wallin and Loonan (1971) observed that heavy rainfall after migration may have reduced the BYDV incidence, and that there was an apparent correlation between temperature following aphid arrival and severity of BYDV.

Secondary spread of BYDV is affected by vector activity, efficiency, and species (Dean, 1973; Halbert and Pike, 1985; Plumb and Lennon, 1982). Since many BYDV isolates are vector specific, it follows that the appropriate vector must be active in fields containing virus-infected plants in order for within-field spread to occur. Patches of infection, apparently due to spread from initial foci, are not uncommon. Additionally, infected plants are more likely to be infested with aphids than are healthy plants, and at least some aphid species reproduce more rapidly and produce more winged individuals on BYDV-infected plants (Gildow, 1980). Random types of distribution are also observed, particularly with spring-sown crops (Dean, 1973; Plumb, 1977). Thus the epidemiology of BYDV is directly related to patterns of aphid migration, movement and overwintering, and the role of reservoir hosts.

For nonpersistent viruses, the low probability of serial transmission makes it likely that an incoming viruliferous vector acquired virus from the last plant it visited. Consequently, spread occurs mainly in a single movement. This contrasts sharply with the situation for persistent viruses, where a vector can retain virus after numerous takeoffs and landings on uninfected plants. In the absence of nonstop transport, the long-distance spread of nonpersistent viruses is unlikely, since vectors that make an intermediate stop on a nonsource plant after acquiring virus usually lose inoculativity. Additionally, the probability of between-field spread is limited by the short nonfeeding retention times of nonpersistent viruses. However, just how far a nonpersistent virus is estimated to be able to spread is highly dependent on the conditions under which retention time is measured.

There are various ways in which aphids carrying a nonpersistent virus can leave the area where infected source plants are located. Their low flight speeds make directed flight difficult except for short distances under low-wind conditions (Johnson, 1969; Pedgely, 1982). However, once an aphid is airborne, it can easily be moved short or long distances by the wind. If an aphid lifts off from a source plant in a moderate wind, it can be transported for moderate distances at low altitude. Under such conditions, once an aphid flies downward or stops actively flying, it will quickly reach a plant. This sort of low-level dispersal can result in a disease gradient extending several miles-downwind from the infection source (Thresh 1976, 1983a). If updrafts carry it higher than a few hundred feet, an aphid may be caught up in rapidly moving low-level jet winds (up to 80 km/h). Such winds can carry aphids for distances of over 1000 km (e.g., Australia to New Zealand; Close and Tomlinson, 1975) and can result in most of a vector population traveling nonstop for a long distance. The environment and behavior of aphids during such long-range dispersal are very different from those under which retention times are usually determined.

An example of likely long-range spread of a nonpersistent virus is the 1977 epidemic of maize dwarf mosaic virus (MDMV) in the northern Great Plains of North America. With the exception of virus retention time, all of Thresh's requirements for long-range spread were met (Thresh, 1983b; Zeyen et al., 1987). That is, there were suitable sources of virus and vectors in the southern Great Plains, the vectors were suitably mobile (alates, via low-level jet winds), vectors were likely to be inoculative soon after takeoff (as is typical for a nonpersistent virus), there were appropriate climatological conditions (warm, sustained, low-level jet winds), and large acreages of highly susceptible, young sweet corn provided an available "target." Additionally, seed transmission of MDMV, which can occur at a low level in sweet corn, was deemed an unlikely explanation for the sudden and massive epiphytotic. Considering possible MDMV source areas, it was estimated that aphids would have had to remain aloft for at least 8 h, and more realistically, for more than 14 h. Thus if the retention of MDMV in aphids exceeded the previously reported maximum time of 6 h, long-distance dispersal was the most plausible explanation for the epidemic (Zeyen et al., 1987). By designing experiments meant to mimic conditions that aphids might experience in a low-level jet wind and by testing thousands of aphids (Berger and Zeyen, 1987; Berger et al., 1987), it was observed that for an aphid population originally transmitting MDMV at an efficiency about 20%, vector efficiency was over 3% after 24 h. The key factor in obtaining such long retention times was denying aphids the opportunity to probe on solid surfaces: a condition not normally achieved in routine laboratory retention time measurement procedures, but the normal condition of aphids during transport in jet winds. The common method for determining retention of inoculativity of nonpersistently transmitted viruses is to place aphids in a suitable container until it is time for them to be assayed. This leads to unnaturally short retention times and does not address the conditions that aphids experience when flying or during low-level jet wind dispersal. When aphids are allowed to probe on solid surfaces or on plants, the rate of loss of inoculativity is significantly greater than that of nonprobing aphids (Berger et al., 1987).

Other considerations are also important in estimating the potential of nonpersistent viruses for long-distance spread. Many reported retention times are based on relatively small samples of insects—sometimes less than 10 at each sampling time, and usually less than 100. If 100 aphids per sample are tested, and the true percentage of aphids that transmit is 0.01% (1 in 10,000 aphids), no transmission will be recorded in 99.0% of such samples (assuming a binomial distribution). Although a 0.01% incidence of viruliferous vectors seems small, it can represent a large number of viruliferous vectors when the enormous size of moving aphid populations is considered. For example, Dickson (1959), in California, calculated that 2.4×10^{10} aphids passed a 1-mile front over a 3-h period. Even if only an extremely small percentage of those aphids were viruliferous after landing, a relatively high incidence of disease could follow. Similarly, if an influx of vectors resulted in an average of one vector per plant landing at random (Hull, 1968), a disease incidence of approximately 0.1% would be expected from a 0.1% incidence of virus in the vector population. If the apparently exponential rate of loss of inoculativity that has been observed at relatively short retention-time intervals also applies to longer times, retention time for some vectors in a population is theoretically infinite and limited only by the longevity of the vectors during transport (Berger, 1983). Observed maximum retention times are thus probably as much a measure of our ability to test large numbers of vectors as they are a measure of the actual maximum retention time. Until more is known about the environmental conditions experienced by vectors during long-range transport and the effects of those conditions on viruliferous aphids, long-range spread must be considered to be a possibility for most nonpersistent viruses.

CONTROLS, FORECASTS, AND MODELS

Transmission characteristics have a bearing on many practices used to control phytoarbovirus diseases. The greatest difference in effects of transmission type on the efficiency of control measures is between the nonpersistent viruses and the other three types. Where there are geographically distinct areas serving as sources of viruliferous vectors (such as overwintering sugar beet fields), introduction of virus by short- to intermediate-range spread can be lessened by locating production fields a suitable distance from the source area (Duffus, 1963; Nelson and Tuttle, 1969; Shepherd and Hills, 1970). Such isolation distances are usually shorter for nonpersistent viruses because they do not usually spread as far as the other types. Also, the relatively low rate of serial transmission of nonpersistent viruses makes them amenable to control using barrier crops (Broadbent, 1957; Jayasena and Randles, 1985; Simons, 1957, 1960). Viruliferous aphids probing on nonhost barrier crops often lose the virus before they probe on susceptible production crops. Persistent and semipersistent viruses can usually be serially transmitted at a much higher rate; consequently, they are not lost by their vectors under similar conditions.

Insecticidal control of vector populations can result in effective control of the spread of some semipersistent and persistent viruses, but not usually of nonpersistent viruses (Bennett, 1971; Hull and Heathcote, 1967; Jayasena and Randles, 1985). This difference is probably due to vectors of nonpersistent viruses being able to acquire and/or transmit in such short periods that incoming vectors can transmit virus before they are killed by the insecticide. In some cases, insecticides may even increase the incidence of nonpersistent viruses (Loebenstein and Raccah, 1980), possibly by increasing the frequency with which vectors probe. The few cases where the spread of a nonpersistent virus has been reported to be controlled by insecticides may have been situations where few new vectors entered the system after the insecticide reduced or eliminated the original vector population.

Because they are not usually subject to insecticidal control, nonpersistent viruses have been the target of a number of novel approaches to control. In contrast to results with insecticides, nonpersistent viruses (and some semipersistent viruses) can sometimes be controlled by sprays of light mineral oil, but such sprays have little effect on persistent viruses (Vanderveken, 1977). This difference may be due to the interference of oils with the unique manner in which virus and vector are associated in nonpersistent transmission. Some control measures aimed at the behavior of incoming aphids would probably be as effective against persistent viruses, but are rarely used if insecticides can be utilized. Uses of reflective mulches, aluminum foil, white nets, border plantings of trap crops, and aphid alarm pheromones are all directed at diminishing the vectors' abilities to detect host plants, and thus lessening the number of potential vectors landing on plants (Cohen, 1982; Cohen and Melamed-Madjar, 1978; Harpaz, 1982; Loebenstein and Raccah, 1980; Raccah, 1986; Simons, 1982). At the present time, breeding disease- or vector-resistant plants is still the most effective means of obtaining control.

Although many of the control measures mentioned above can be effective and are used in specific situations, cost-effectiveness limits their more widespread use. For many diseases, potentially effective controls are often not implemented because growers are not sure that a particular disease will appear in their crops. A control that is used without disease pressure being present is often considered wasted.

For a few diseases it may be possible to predict the occurrence of large vector populations far enough in advance to allow growers to implement a control measure soon enough to be effective. One of the few examples of such a forecasting system is the use of multifield aphid counts made by processing plant employees to predict the need for insecticide applications to control sugarbeet yellows and beet mild yellows viruses in England (Hull, 1968; Taylor, 1986). This forecasting system could potentially be made more accurate through incorporation of weather data (Watson et al., 1975). Such regional prediction is often impossible because there are too many potential vector sources (Annand et al., 1932; Kisimoto, 1976; Rosenberg and Magor, 1983). The use of vector counts to predict particular levels of disease in particular production fields is not usually economically feasible because of the relatively large amounts of field data necessary to characterize a particular vector population (Madden et al., 1983).

Although explicit forecasting of particular vector influxes has rarely been used on an economic scale, many common recommendations for control of arthropod-borne viruses are based on implicit forecasts of disease. For example: the forecast "if sugar beets are planted adjacent to an overwintering beet field, a high incidence of beet yellows will probably occur" implies a recommendation that production fields should be isolated from overwintering seed crops. Such generalized forecast statements have been the main practical results of the few mathematical models of virus–vector–plant relationships that have been developed (Allen, 1983; Fishman et al., 1983; Frazer, 1977; Kiritani and Sasaba, 1978; Marcus and Raccah, 1986; Ruesink and Irwin, 1986; Watson and Healy, 1953).

Both simple algebraic models (Allen, 1983) and more complex computer simulations (Fishman et al., 1983) have been used successfully to predict the effects of removal of infected plants on the spread of disease, and thus to determine whether such roguing might be economically justified. Additionally, models can often aid in the development of useful hypotheses about disease processes; such as the finding that ladybugs (Coleoptera: Coccinellidae) may increase alfafa mosaic virus by inducing vectors to move more frequently, even though predation by ladybugs reduces the size of the vector population (Frazer, 1977). Such direct practical uses of existent models warrant the more frequent use of modeling approaches in the study of virus disease epidemiology. However, we believe that the greatest potential value of such models is as indirect aids in the development of an understanding of disease behavior, rather than in specific predictions. The models of disease behavior that have the greatest impact on the practical control of disease are those that people construct mentally rather than those represented by computer programs or sets of equations. Both concrete and theoretical research on the epidemiology and transmission mechanisms of phytoarbovirus diseases must ultimately be directed at developing these conceptual, but most basic, "models."

ACKNOWLEDGMENTS

The authors wish to thank Drs. B. Raccah, T. P. Pirone, and R. J. Zeyen for their helpful comments and suggestions.

REFERENCES

ALLEN, R. N. 1983. Spread of banana bunchy top and other plant virus diseases in time and space. *In* Plant Virus Epidemiology (ed. R. T. Plumb and J. T. Thresh), pp. 51–59. Blackwell Scientific Publications Ltd., Oxford.

ANANTHAKRISHNAN, T. N. 1980. Thrips. *In* Vectors of Plant Pathogens (ed. K. F. Harris and K. Maramorosch), pp. 149–164. Academic Press, Inc., New York.

ANNAND, P. N., CHAMBERLAIN, J. C., HENDERSON, C. F., and WATERS, H. A. 1932. Movements of the beet leafhopper in 1930 in southern Idaho. USDA Circ. 244.

ARMOUR, S. L., MELCHER, V., PIRONE, T. P., and ESSENBERG, R. C. 1983. Helper component for aphid transmission encoded by region II of cauliflower mosaic virus DNA. Virology 129:25–30.

BANCROFT, J. B., ULLSTRUP, A. J., MESSIEHA, M., BRACKER, C. E., and SNAZELLE, T. E. 1966. Some biological and physical properties of a midwestern isolate of maize dwarf mosaic virus. Phytopathology 56:474–478.

BANTTARI, E. E., and ZEYEN, R. J. 1970. Transmission of oat blue dwarf virus by the aster leafhopper following natural acquisition or inoculation. Phytopathology 60:399–402.

BANTTARI, E. E., and ZEYEN, R. J. 1973. Oat blue dwarf virus. Commonwealth Mycological Institute/Association of Applied Biologists. Descriptions of Plant Viruses, No. 123.

BAR-JOSEPH, M., and MURANT, A. F. 1982. Closterovirus group. Commonwealth Mycological Institute/Association of Applied Biologists. Descriptions of Plant Viruses, No. 260.

BAR-JOSEPH, M., GARNSEY, S. M., and GONSALVES, D. 1979. The closteroviruses: a distinct group of elongated plant viruses. Adv. Virus Res. 25:93–168.

BAWDEN, F. C., and KASSANIS, B. 1946. Varietal differences in susceptibility to potato virus Y. Ann. Appl. Biol. 33:46–50.

BAWDEN, F. C., and KASSANIS, B. 1950. Some effects of host nutrition on the susceptibility of plants to infection by certain viruses. Ann. Appl. Biol. 37:46–57.

BENNETT, C. W. 1967. Epidemiology of leafhopper-transmitted viruses. Annu. Rev. Phytopathol. 5:87–108.

BENNETT, C. W. 1971. The curly top disease of sugarbeet and other plants. Am. Phytopathol. Soc. Monogr. 7.

BENNETT, C. W., and WALLACE H. E. 1938. Relationship of the curly-top virus to the vector, *Eutettix tenellus.* J. Agric. Res. 56:31–51.

BERGER, P. H. 1983. The retention of maize dwarf mosaic virus by the greenbug, *Schizaphis graminum* Rondani and its implications for transmission mechanisms. Ph.D. dissertation, Texas A&M University.

BERGER, P. H., and PIRONE, T. P. 1986. The effect of helper component on uptake and localization of potyviruses in *Myzus persicae.* Virology 153:256–261.

BERGER, P. H., and ZEYEN, R. J. 1987. Effects of sustained immobilization on aphids. Ann. Appl. Biol. 111:247–256.

BERGER, P. H., TOLER, R W., and HARRIS, K. F. 1983. Maize dwarf mosaic virus transmission by greenbug biotypes. Plant Dis. 67:496–497.

BERGER, P. H., ZEYEN, R. J., and GROTH, J. V. 1987. Aphid retention of maize dwarf mosaic virus: Epidemiological implications. Ann. Appl. Biol. 111:337–344.

BIRD, J., and MARAMOROSCH, K. 1978. Viruses and diseases associated with whiteflies. Adv. Virus Res. 22:55–110.

BLACK, L. M. 1959. Biological cycles of plant viruses in insect vectors. *In* The Viruses, Vol. 2 (ed. F. M. Burnet and W. M. Stanley), pp. 157–185. Academic Press, Inc., New York.

BOCCARDO G., and MILNE, R. G. 1984. Plant reovirus group. Commonwealth Mycological Institute/Association of Applied Biologists. Descriptions of Plant Viruses, No. 294.

BOS, L., 1981. Wild plants in the ecology of virus diseases. *In* Plant Diseases and Vectors (ed. K. Maramorosch and K. F. Harris), pp. 1–33. Academic Press, Inc., New York.

BRADLEY, R. H. E. 1964. Aphid transmission of stylet-borne viruses. *In* Plant Virology (ed. M. K. Corbett and H. D. Sisler), pp. 148–174. The University of Florida Libraries, Gainseville, Fla.

BRADLEY, R. H. E., and GANONG, R. Y. 1955a. Evidence that potato virus Y is carried near the tips of the stylets of the aphid vector *Myzus persicae* (Sulz.). Can. J. Microbiol. 1:775–782.

BRADLEY, R. H. E., and GANONG, R. Y. 1955b. Some effects of formaldehyde on potato virus Y *in vitro,* and ability of aphids to transmit the virus when their stylets are treated with formaldehyde. Can. J. Microbiol. 1:783–798.

BRADLEY, R. H. E., and RIDEOUT, D. W. 1953. Comparative transmission of potato virus Y by four aphid species that infest potato. Can. J. Zool. 31:333–341.

BROADBENT, L. 1953. Aphids and virus diseases in potato crops. Biol. Rev. 28:350–380.

BROADBENT, L. 1957. Investigation of virus diseases of brassica crops. A.R.C. Rep. Ser. 14. Cambridge University Press, Cambridge.

BRUEHL, G. W. 1961. Barley yellow dwarf. Am. Phytopathol. Soc. Monogr. 1.

BRUNT, A. A. 1970. Cacao swollen shoot virus. Commonwealth Mycological Institute/Association of Applied Biologists. Descriptions of Plant Viruses, No. 10.

BRYANT, E. 1969. A model of plant-to-plant movement of aphids: a new approach. Res. Popul. Ecol. (Kyoto) 11:34–39.

CAMPBELL, C. L., and NOE, J. P. 1985. The spatial analysis of soilborne pathogens and root diseases. Annu. Rev. Phytopathol. 23:129–148.

CARTER, N., MCLEAN, I. F. G., WATT, A. D., and DIXON, A. F. G. 1980. Cereal aphids: a case study and review. Appl. Biol. 5:271–348.

CARTIER, J. J. 1963. Varietal resistance of peas to pea aphid biotypes under field and greenhouse conditions. J. Econ. Entomol. 56:205–213.

CHALFANT, R. B., and CHAPMAN, R. K. 1962. Transmission of cabbage viruses A and B by the cabbage aphid and the green peach aphid. J. Econ. Entomol. 55:584–590.

CHIU, R.-J., JEAN, J. H., CHEN, M. H., and LO, T. C. 1968. Transmission of transitory yellowing of rice by two leafhoppers. Phytopathology 58:740–745.

CHIYKOWSKI, L. N. 1953. Studies on migration and control of the six-spotted leafhopper, *Macrosteles fascifrons* (Stål), in relation to the transmission of aster yellows virus. Ph.D. thesis, University of Wisconsin.

CHIYKOWSKI, L. N. 1981. Epidemiology of diseases caused by leafhopper-borne pathogens. *In* Plant Diseases and Vectors (ed. K. Maramorosch and K. F. Harris), pp. 105–159. Academic Press, Inc., New York.

CHIYKOWSKI, L. N., and CHAPMAN, R. K. 1965. Migration of the six-spotted leafhopper *Macrosteles fascifrons* (Stål). Part 2. Migration of the six-spotted leafhopper in central North America. Res. Bull. Agric. Exp. Stn. Univ. Wis. 261:21–45.

CLOSE, R. L., and TOMLINSON, A. I. 1975. Dispersal of the grain aphid *Macrosiphum miscanthi* from Australia to New Zealand. N.Z. Entomol. 6:62–65.

COHEN, S. 1982. Control of whitefly vectors by color mulches. *In* Pathogens, Vectors, and Plant Diseases (ed. K. F. Harris and K. Maramorosch), pp. 45–56. Academic Press, Inc., New York.

COHEN, S., and MELAMED-MADJAR, V. 1978. Prevention by soil mulching of the spread of tomato yellow leaf curl virus transmitted by *Bemisia tabaci* (Gennadius) (Hemiptera: Aleyrodidae) in Israel. Bull. Entomol. Res. 68:465–470.

CONIN, R. V. 1956. Oversummering volunteer wheat in the epidemiology of wheat streak mosaic virus. J. Econ. Entomol. 49:405–406.

CONTI, M. 1985. Transmission of plant viruses by leafhoppers and planthoppers. *In* The Leafhoppers and Planthoppers (ed. L. R. Nault and J. G. Rodriguez), pp. 289–307. John Wiley & Sons, Inc., New York.

COOK, W. C. 1967. Life history, host plants, and migration of the beet leafhopper in the western United States. USDA Tech. Bull. 1365. pp. 1–122.

COSTA, A. S. 1969. White flies as virus vectors. *In* Viruses, Vectors, and Vegetation (ed. K. Maramorosch), pp. 95–119. John Wiley & Sons, Inc., New York.

COSTA, A. S. 1975. Increase in the population density of *Bemisia tabaci,* a threat of widespread infection of legume crops in Brazil. *In* Tropical Diseases of Legumes (ed. J. Bird and K. Maramorosch), pp. 27–49. Academic Press, Inc., New York.

DALE, W. T. 1953. The transmission of plant viruses by biting insects, with particular reference to cowpea mosaic virus. Ann. Appl. Biol. 40:385–392.

DEAN, G. J. W. 1973. Distribution of aphids in spring cereals. J. Econ. Entomol. 52:994–996.

DEAN, G. J. W. 1974. Overwintering and abundance of cereal aphids. Ann. Appl. Biol. 76:1–7.

DELONG, D. M. 1971. The bionomics of leafhoppers. Annu. Rev. Entomol. 16:179–210.

DICKSON, R. C. 1959. Aphid dispersal over southern California deserts. Ann. Entomol. Soc. Am. 52:368–372.

DRAKE, D. C., and CHAPMAN, R. K. 1965. Migration of the six-spotted leafhopper *Macrosteles fascifrons* (Stål). Part 1. Evidence for long-distance migration of the six-spotted leafhopper into Wisconsin. Res. Bull. Agric. Exp. Stn. Univ. Wis. 261: 3–20.

DUFFUS, J. E. 1963. Incidence of beet virus diseases in relation to overwintering beet fields. Plant Dis. Rep. 47:428–431.

DUFFUS, J. E. 1977. Aphids, viruses, and the yellow plague. *In* Aphids as Virus Vectors (ed. K. F. Harris and K. Maramorosch), pp. 361–383. Academic Press, Inc., New York.

EASTOP, V. F. 1977. Worldwide importance of aphids as virus vectors. *In* Aphids as Virus Vectors (ed. K. F. Harris and K. Maramorosch), pp. 385–412. Academic Press, Inc., New York.

ELTON, C. 1925. The dispersal of insects to Spitzbergen. Trans. R. Entomol. Soc. London 1925:289–299.

ENGELBRECHT, D. J., and KASDORF, G. G. F. 1985. Association of a closterovirus with grapevines indexing positive for grapevine leafroll disease and evidence for its natural spread in grapevine. Phytopathol. Mediterr. 24:101–105.

ENTWISTLE, P. F. 1973. Coccoids. *In* Viruses and Invertebrates (ed. A. J. Gibbs), p. 178–191. Elsevier/North-Holland, Amsterdam.

EVERETT, T. R., and LAMEY, H. A. 1969. Hoja blanca. *In* Viruses, Vectors, and Vegetation (ed. K. Maramorosch), pp. 361–377. John Wiley & Sons, Inc., New York.

FINE, P. E. M., and SYLVESTER, E. S. 1978. Calculation of vertical transmission rates of infection, illustration with data on an aphid-borne virus. Am. Nat. 112:781–786.

FISHMAN, S., MARCUS, R., TALPAZ, H., BAR-JOSEPH, M., OREN, Y., SALOMON, R., and ZOHAR, M. 1983. Epidemiological and economic models for spread and control of citrus tristeza virus disease. Phytoparasitica 11:39–49.

FLEMING, R. A., MARSH, L. M., and TUCKWELL, H. C. 1982. Effect of field geometry on the spread of crop disease. Prot. Ecol. 4:81–108.

FORBES, A. R., and MACCARTHY, H. R. 1969. Morphology of the Homoptera, with emphasis on virus vectors. *In* Viruses, Vectors and Vegetation (ed. K. Maramorosch), pp. 211–234. John Wiley & Sons, Inc., New York.

FRANCKI, R. I. B., and HATTA, T. 1981. Tomato spotted wilt virus. *In* Handbook of Plant Virus Infections (ed. E. Kurstak), pp. 491–511. Elsevier/North-Holland, Amsterdam.

FRANCKI, R. I. B., KATAJIMA, E. W., and PETERS, D. 1981. Rhabdoviruses. *In* Handbook of Plant Virus Infections (ed. E. Kurstak), pp. 455–489. Elsevier/North-Holland, Amsterdam.

FRAZER, B. D. 1977. Plant virus epidemiology and computer simulation of aphid populations. *In* Aphids as Virus Vectors (ed. K. F. Harris and K. Maramorosch), pp. 413–431. Academic Press, Inc., New York.

FREITAG, J. H. 1941a. Insect transmission, host range, and properties of squash mosaic virus. Phytopathology (Abstr.) 31:8.

FREITAG, J. H. 1941b. A comparison of the transmission of four curcurbit viruses by cucumber beetles and by aphids. Phytopathology (Abstr.) 31:8.

FREITAG, J. H. 1956. Beetle transmission, host range, and properties of squash mosaic virus. Phytopathology 46:73–81.

FUKUSHI, T. 1969. Relationships between propagative rice viruses and their vectors. *In* Viruses, Vectors, and Vegetation (ed. K. Maramorosch), pp. 279–301. John Wiley & Sons, Inc., New York.

FULTON, J. P., SCOTT, H. A., and GAMEZ, R. 1975. Beetle transmission of legume viruses. *In* Diseases of Tropical Legumes (ed. J. Bird and K. Maramorosch), pp. 121–131. Academic Press, Inc., New York.

FULTON, J. P., SCOTT, H. A., and GAMEZ, R. 1980. Beetles. *In* Vectors of Plant Pathogens (ed. K. F. Harris and K. Maramorosch), pp. 115–132. Academic Press, Inc., New York.

GALVEZ, G. E. 1971. Rice tungro virus. Commonwealth Mycological Institute/Association of Applied Biologists. Descriptions of Plant Viruses, No. 67.

GAMEZ, R. 1980. Maize rayado fino virus. Commonwealth Mycological Institute/Association of Applied Biologists. Descriptions of Plant Viruses, No. 220.

GAMEZ, R., and WATSON, M. A. 1964. Failure of anesthetized aphids to acquire or transmit henbane mosaic virus when their stylets were artificially inserted into leaves of infected or healthy tobacco. Virology 22:292–295.

GARRETT, R. G. 1973. Non-persistent aphid-borne viruses. *In* Viruses and Invertebrates (ed. A. J. Gibbs), pp. 476–492. North-Holland Publishing Company, Amsterdam.

GATES, L. F. 1970. The potential of corn and wheat to perpetuate wheat streak mosaic virus in southern Ontario. Can. Plant Dis. Surv. 50:59–62.

GERA, A., LOEBENSTEIN, G., and RACCAH, B. 1979. Protein coats of two strains of cucumber mosaic virus affect aphid transmission by *Aphis gossypii.* Phytopathology 69:396–399.

GERGERICH, R. C., SCOTT, H. A., and FULTON, J. P. 1983. Regurgitant as a determinant of specificity in the transmission of plant viruses by beetles. Phytopathology 73:936–938.

GIBSON, W. W., and PAINTER, R. H. 1957. Transport by aphids of the wheat curl mite, *Aceria tulipae* (K.), a vector of wheat streak mosaic virus. J. Kans. Entomol. Soc. 30:147–153.

GILBERT, N. 1982. Comparative dynamics of a single-host aphid. III. Movement and population structure. J. Anim. Ecol. 51:469–480.

GILDOW, F. E. 1980. Increased production of alatae by aphids reared on oats infected with barley yellow dwarf virus. Ann. Entomol. Soc. Am. 73:343–347.

GILDOW, F. E. 1982. Coated vesicle transport of luteoviruses through salivary glands of *Myzus persicae.* Phytopathology 72:1289–1296.

GILDOW, F. E. 1985. Transcellular transport of barley yellow dwarf virus into the hemocoel of the aphid vector, *Rhopalosiphum padi.* Phytopathology 75:292–297.

GILDOW, F. E., and ROCHOW, W. F. 1980. Role of accessory salivary glands in aphid transmission of barley yellow dwarf virus. Virology 104:97–108.

GINGERY, R. E., BRADFUTE, O. E., GORDON, D. T., and NAULT, L. R. 1978. Maize chlorotic dwarf virus. Commonwealth Mycological Institute/Association of Applied Biologists. Descriptions of Plant Viruses, No. 194.

GINGERY, R. E., GORDON, D. T., NAULT, L. R., and BRADFUTE, O. E. 1981. Maize chlorotic dwarf virus. *In* Handbook of Plant Virus Infections (ed. E. Kurstak), pp. 19–32. Elsevier/North-Holland, Amsterdam.

GLICK, P. A. 1960. Collecting insects by airplane, with special reference to dispersal of the potato leafhopper. USDA Tech. Bull. 1222.

GOODMAN, R. M. 1977. Single stranded DNA genome in a whitefly-transmitted plant virus. Virology 83:171–179.

GOODMAN, R. M., BIRD, J., and THONGMEEARKOM, P. 1977. An unusual virus-like particle associated with golden yellow mosaic of beans. Phytopathology 67:37–42.

GOVIER, D. A., and KASSANIS, B. 1974. A virus-induced component of plant sap needed when aphids acquire potato virus Y from purified preparations. Virology 61:420–426.

GRANADOS, R. R. 1969. Maize viruses and vectors. *In* Viruses, Vectors, and Vegetation (ed. K. Maramorosch), pp. 327–359. John Wiley & Sons, Inc., New York.

GRAY, S. M., MOYER, J. W., and BLOOMFIELD, P. 1986. Two-dimensional distance class model for quantitative description of virus-infected plant distribution lattices. Phytopathology 76:243–248.

GREGORY, P. H. 1968. Interpreting plant disease dispersal gradients. Annu. Rev. Phytopathol. 6:189–212.

GUTIERREZ, A. P., MORGAN, D. J., and HAVENSTEIN, D. E. 1971. The ecology of *Aphis craccivora* Koch and subterraneum clover stunt virus. I. The phenology of aphid populations and the epidemiology of virus in pastures in south-east Australia. J. Appl. Ecol. 8:699–721.

GUTIERREZ, A. P., HAVENSTEIN, D. E., NIX, H. A., and MOORE, P. H. 1974. The ecology of *Aphis craccivora* and subterranean clover stunt virus in south-east Australia. II. A model of cowpea aphid populations in temperate pastures. J. Appl. Ecol. 11:1–20.

GUTIERREZ, A. P., BAUMGAERTNER, J. U., and SUMMERS, C. G. 1984. Multitropic models of predator–prey energetics. Can. Entomol. 116:923–963.

HALBERT, S. E., and PIKE, K. S. 1985. Spread of barley yellow dwarf virus and relative importance of local aphid vectors in central Washington. Ann. Appl. Biol. 107:387–395.

HARPAZ, I. 1972. Maize Rough Dwarf. Israel Universities Press, Jerusalem.

HARPAZ, I. 1982. Nonpesticidal control of vector-borne diseases. *In* Pathogens, Vectors, and Plant Diseases (ed. K. F. Harris and K. Maramorosch), pp. 1–21. Academic Press, Inc., New York.

HARREWIJN, P., VAN HOOF, H. A., and NOORDINK, J. P. W. 1981. Flight behavior of the aphid *Myzus persicae* during its maiden flight. Neth. J. Plant Pathol. 87:111–117.

HARRIS, K. F. 1977. An ingestion–egestion hypothesis of noncirculative virus transmission. *In* Aphids as Virus Vectors (ed. K. F. Harris and K. Maramorosch), pp. 165–220. Academic Press, Inc., New York.

HARRIS, K. F. 1978. Aphid-borne viruses: ecological and environmental aspects. *In* Viruses and Environment (ed. E. Kurstak and K. Maramorosch), pp. 311–337. Academic Press, Inc., New York.

HARRIS, K. F. 1979. Leafhopper and aphids as biological vectors: vector–virus relationships. *In* Leafhopper Vectors and Plant Disease Agents (ed. K. Maramorosch and K. F. Harris), pp. 217–308. Academic Press, Inc., New York.

HARRIS, K. F. 1981. Arthropod and nematode vectors of plant viruses. Annu. Rev. Phytopathol. 19:391–426.

HARRIS, K. F. 1983. Sternorrhynchous vectors of plant viruses: virus–vector interactions and transmission mechanisms. Adv. Virus Res. 28:113–140.

HARRIS, K. F., and BATH, J. E. 1973. Regurgitation by *Myzus persicae* during membrane feeding: its likely function in transmission of nonpersistent plant viruses. Ann. Entomol. Soc. Am. 66:793–796.

HARRIS, K. F., BATH, J. E., THOTTAPPILLY, G., and HOOPER, G. R. 1975. Fate of pea enation mosaic virus in PEMV-injected pea aphids. Virology 65:148–162.

HARRISON, B. D. 1977. Ecology and control of viruses with soil-inhabiting vectors. Annu. Rev. Phytopathol. 14:331-360.

HARRISON, B. D. 1981. Plant virus ecology: ingredients, interactions and environmental influences. Ann. Appl. Biol. 99:195–209.

HARRISON, B. D., and MURANT, A. F. 1985. Descriptions of Plant Viruses. CMI/AAB, Slough, England.

HODSON, A. C., and COOK, E. F. 1960. Long-range aerial transport of the harlequin bug and greenbug into Minnesota. J. Econ. Entomol. 53:604–608.

HOLZAPFEL, E. P., TSUDA, D. M., and HARREL, J. 1970. Trapping of air-borne insects in the antarctic area (Part 3). Pac. Insects 12:133–156.

HUFF, F. H. 1963. Relation between leafhopper influxes and synoptic weather conditions. J. Appl. Meteorol. 2:39–43.

HULL, R. 1968. The spray warning scheme for control of sugar beet yellows in England. Summary of results between 1959–66. Plant Pathol. 17:1–10.

HULL, R., and HEATHCOTE, G. D. 1967. Experiments on the time of application of insecticide to decrease the spread of yellowing viruses of sugar beet, 1954–66. Ann. Appl. Biol. 60:469–478.

IE, T. S. 1970. Tomato spotted wilt virus. Commonwealth Mycological Institute/Association of Applied Biologists. Descriptions of Plant Viruses, No. 39.

IRWIN, M. E., and GOODMAN, R. M. 1981. Ecology and control of soybean mosaic virus. *In* Plant Diseases and Vectors: Ecology and Epidemiology (ed. K. Maramorosch and K. F. Harris), pp. 181–220. Academic Press, Inc., New York.

IRWIN, M. E., and RUESINK, W. G. 1986. Vector intensity: a product of propensity and activity. *In* Plant Virus Epidemics: Monitoring, Modeling, and Predicting Outbreaks (ed. G. D. McLean, R. G. Garrett, and W. G. Ruesink), pp. 13–33. Academic Press, Sydney.

JAYASENA, K. W., and RANDLES, J. W. 1985. The effect of insecticides and a plant barrier row on aphid populations and the spread of bean yellow mosaic potyvirus and subterranean clover red leaf luteovirus in *Vicia faba* in south Australia. Ann. Appl. Biol. 107:355–364.

JEGER, M. J. 1983. Analyzing epidemics in time and space. Plant Pathol. 35:5–12.

JOHNSON, B. 1957. Studies on the dispersal by upperwinds of *Aphis craccivora* Koch in New South Wales. Proc. Linn. Soc. N.S.W. 82:191–198.

JOHNSON, C. G. 1967. International dispersal of insects and insect-borne viruses. Neth. J. Plant Pathol. 73(Suppl. 1):21–43.

JOHNSON, C. G. 1969. Migration and Dispersal of Insects by Flight. Methuen & Company, Ltd., London.

KAMPMEIJER, P., and ZADOKS, J. C. 1977. EPIMUL, a Simulator of Foci and Epidemics in Mixtures of Resistant and Susceptible plants, Mosaics and Multilines. PUDOC Wageningen, The Netherlands.

KENNEDY, J. S., BOOTH, C. O., and KERSHAW, W. J. 1959. Host finding by aphids in the field. Ann. Appl. Biol. 47:424–444.

KENNEDY, J. S., DAY, M. F., and EASTOP, V. F. 1962. A Conspectus of Aphids as Vectors of Plant Viruses. Commonwealth Institute of Entomology, London.

KIRITANI, K., and SASABA, T. 1978. An experimental validation of the system model for prediction of rice dwarf virus infection. Appl. Entomol. Zool. 13:209–214.

KISIMOTO, R. 1967. Genetic variation in the ability of a planthopper vector; *Laodelphax striatellus* (Fallen) to acquire the rice stripe virus. Virology 32:144–152.

KISIMOTO, R. 1973. Leafhoppers and planthoppers. *In* Viruses and Invertebrates (ed. A. J. Gibbs), pp. 137–156. North-Holland Publishing Company, Amsterdam.

KISIMOTO, R. 1976. Synoptic weather conditions inducing long-distance immigration of planthoppers, *Sogatella furcifera* Horvath and *Nilaparvata luqens* Stål. Ecol. Entomol. 1:95–109.

KNOKE, J. K., and LOUIE, R. 1981. Epiphytology of maize virus diseases. South. Coop. Ser. Bull. 247:92–102.

LAIRD, E. F., JR., and DICKSON, R. C. 1963. Tobacco etch virus and potato virus Y in pepper, their host plants and insect vectors in southern California. Phytopathology 53:48–52.

LIM, W. L., and HAGEDORN, D. J. 1977. Bimodal transmission of plant viruses. *In* Aphids as Virus Vectors (ed. K. F. Harris and K. Maramorosch), pp. 237–251. Academic Press, Inc., New York.

LING, K. C. 1966. Nonpersistence of the tungro virus in its leafhopper vector, *Nephotettix impicticeps.* Phytopathology 56:1252–1256.

LING, K. C. 1969. Nonpropagative leafhopper-borne viruses. *In* Viruses, Vectors, and Vegetation (ed. K. Maramorosch) pp. 255–277. John Wiley & Sons, Inc., New York.

LING, K. C., and TIONGCO, E. R. 1979. Transmission of rice tungro virus at various temperatures: a transitory virus–vector interaction. *In* Leafhopper Vectors and Plant Disease Agents (ed. K. Maramorosch and K. F. Harris), pp. 349–366. Academic Press, Inc., New York.

LISTER, R. M., and BAR-JOSEPH, M. 1981. Closteroviruses. *In* Handbook of Plant Virus Infections (ed. K. Kurstak), pp. 809–844. Elsevier/North-Holland, New York.

LOEBENSTEIN, G., and RACCAH, B. 1980. Control of nonpersistently transmitted aphid-borne viruses. Phytoparasitica 8:221–235.

LUNG, M. C. Y., and PIRONE, T. P. 1973. Studies on the reason for differential transmissibility of cauliflower mosaic virus isolates by aphids. Phytopathology 63:910–914.

LUNG, M. C. Y., and PIRONE, T. P. 1974. Acquisition factor required for aphid transmission of purified cauliflower mosaic virus. Virology 60:260–264.

MACKENZIE, D. R., BARFIELD, C. S., KENNEDY, G. G., and BERGER, R. D. (eds.) 1985. The Movement and Dispersal of Agriculturally Important Biotic Agents. Claitor's Publishing Division, Baton Rouge, La.

MADDEN, L. V. 1985. Modeling the population dynamics of leafhoppers. *In* The Leafhoppers and Planthoppers (ed. L. R. Nault and J. G. Rodriguez), pp. 235–258. John Wiley & Sons, Inc., New York.

MADDEN, L. V., and CAMPBELL, C. L. 1986. Description of virus disease epidemics in time and space. *In* Plant Virus Epidemics: Monitoring, Modeling, and Predicting Outbreaks (ed. G. D. McLean, R. G. Garrett, and W. G. Ruesink), pp. 273–293. Academic Press, Inc., New York.

MADDEN, L. V., LOUIE, R., ABT, J. J., and KNOKE, J. K. 1982. Evaluation of tests for randomness of infected plants. Phytopathology 72:195–198.

MADDEN, L. V., KNOKE, J. K., and LOUIE, R. 1983. The statistical relationship between aphid trap catches and maize dwarf mosaic virus inoculation pressure. *In* Plant Virus Epidemiology (ed. R. T. Plumb and J. M. Thresh), pp. 159–168. Blackwell Scientific Publications Ltd., Oxford.

MARAMOROSCH, K., and HARRIS, K. F. 1979. Leafhopper Vectors and Plant Disease Agents. Academic Press, Inc., New York.

MARCUS, R., and RACCAH, B. 1986. A model for spread of non-persistent virus diseases. J. Appl. Stat. (in press).

MATTHEWS, R. E. F. 1982. Classification and nomenclature of viruses. Intervirology 17:1–199.

MCCARTNEY, H. A., and FITT, B. D. L. 1985. Construction of dispersal models. *In* Advances in Plant Pathology, Vol. 3 (ed. C. A. Gilligan), pp. 107–143. Academic Press, Inc., New York.

MCLEAN, G. D., GARRETT, R. G., and RUESINK, W. G. (Eds). 1986. Plant Virus Epidemics: Monitoring, Modeling, and Predicting Outbreaks. Academic Press, Inc., New York.

MEDLER, J. T. 1962. Long-range displacement of Homoptera in the central United States. Proc. 11th Int. Congr. Entomol. 3:30–35.

MEDLER, J. T., and SMITH, P. W. 1960. Greenbug dispersal and distribution of barley yellow dwarf virus in Wisconsin. J. Econ. Entomol. 53:473–474.

MESSIEHA, M. 1967. Aphid transmission of maize dwarf mosaic virus. Phytopathology 57:956–959.

MILNE, R. G., and FRANCKI, R. I. B. 1984. Should tomato spotted wilt be considered as a posible member of the family *Bunyaviridae*? Intervirology 22:72–76.

MILNE, R. G., and LOVISOLO, O. 1977. Maize rough dwarf and related viruses. Adv. Virus Res. 21:267–341.

MONIS, J., SCOTT, H. A., and GERGERICH, R. C. 1986. Effect of beetle regurgitant on plant virus transmission using the gross wounding technique. Phytopathology 76:808–811.

MOSSOP, D. W., and FRANCKI, R. I. B. 1977. Association of RNA 3 with aphid transmission of cucumber mosaic virus. Virology 81:177–181.

MOUND, L. A. 1973. Thrips and whitefly. *In* Viruses and Invertebrates (ed. A. J. Gibbs), pp. 229–242. Van Nostrand Reinhold Company, Inc., New York.

MUNIYAPPA, V. 1980. Whiteflies. *In* Vectors of Plant Pathogens (ed. K. F. Harris and K. Maramorosch), pp. 39–85. Academic Press, Inc., New York.

MURANT, A. F. 1978. Recent studies on association of two plant virus complexes with aphid vectors. *In* Plant Disease Epidemiology (ed. P. R. Scott and A. Bainbridge), pp. 242–249. Blackwell Scientific Publications Ltd., Oxford.

MURANT, A. F., ROBERTS, I. M., and ELNAGAR, S. 1976. Association of virus-like particles with the foregut of the aphid *Cavariella aegopodii* transmitting the semipersistent viruses anthriscus yellows and parsnip yellow fleck. J. Gen. Virol. 31:47–57.

MURANT, A. F., RACCAH, B., and PIRONE, T. P. 1988. Transmission by vectors. *In* The Filamentous Plant Viruses (ed. R. G. Milne), pp. 237–273. Plenum Press Corp., New York.

NAULT, L. R., and BRADLEY, R. H. E. 1969. Acquisition of maize dwarf mosaic virus by the greenbug, *Schizaphis graminum.* Ann. Entomol. Soc. Am. 62:403–406.

NAULT, L. R., and RODRIGUEZ, J. G. 1985. The Leafhoppers and Planthoppers. John Wiley & Sons, Inc., New York.

NAULT, L. R., and STYER, W. E. 1969. The dispersal of *Aceria tulipae* and three other grass-infesting eriophyid mites in Ohio. Ann. Entomol. Soc. Am. 62:1446–1455.

NAULT, L. R., STYER, W. E., KNOKE, J. K., and PITRE, H. N. 1973. Semipersistent transmission of leafhopper-borne maize chlorotic dwarf virus. J. Econ. Entomol. 66:1271–1273.

NELSON, M. R., and TUTTLE, D. M. 1969. The epidemiology of cucumber mosaic and watermelon mosaic 2 on cantaloups in an arid climate. Phytopathology 59:849–856.

NICHIPORICK, W. 1965. The aerial migration of the six-spotted leafhopper and the spread of the virus disease aster yellows. Int. J. Biometeorol. 9:219–227.

NICOT, T. C., ROUSE, D. I., and YANDELL, B. S. 1984. Comparison of statistical methods for the study of spatial patterns of soilborne plant pathogens in the field. Phytopathology 74:1399–1402.

NIELSON, M. W. 1968. The leafhopper vectors of phytopathogenic viruses (Homoptera, Cicadellidae) taxonomy, biology, and virus transmission. USDA Agric. Res. Serv. Tech. Bull. 1382.

NIELSON, M. W. 1979. Taxonomic relationships of leafhopper vectors of plant pathogens. *In* Leafhopper Vectors and Plant Disease Agents (ed. K. Maramorosch and K. F. Harris), pp. 3–27. Academic Press, Inc., New York.

NUORTEVA, P., and HOOGSTRAAL, H. 1963. The incidence of ticks on migrating birds arriving in Finland during the spring of 1962. Ann. Med. Exp. Biol. Fenn. 41:457–468.

OLDFIELD, G. N. 1970. Mite transmission of plant viruses. Annu. Rev. Entomol. 15:343–380.

ORLOB, G. B. 1962. Further studies on the transmission of plant viruses by different forms of aphids. Virology 16:301–304.

ORLOB, G. B. 1966. Feeding and transmission characteristics of *Aceria tulipae* Keifer as a vector of wheat streak mosaic virus. Phytopathol. Z. 55:218–238.

ORLOB, G. B., and ARNY, D. C. 1960. Transmission of barley yellow dwarf virus by different forms of the apple grain aphid, *Rhopalosiphum fitchii* (Sand.). Virology 11:273–274.

OSWALD, J. W., and HOUSTON, B. R. 1951. A new virus disease of cereals, transmissible by aphids. Plant Dis. Rep. 35:471–475.

OWUSU, G. K. 1983. The cocoa swollen shoot disease problem in Ghana. *In* Plant Virus Epidemiology (ed. J. T. Plumb and J. M. Thresh), pp. 73–83. Blackwell Scientific Publications Ltd., Oxford.

PADY, S. M. 1955. The occurrence of the vector of wheat streak mosaic, *Aceria tulipae,* on slides exposed to the air. Plant Dis. Rep. 30:296–297.

PALIWAL, Y. C. 1979. Occurrence and localization of spherical virus-like particles in tissues of apparently healthy tobacco thrips, *Frankliniella fusca,* a vector of tomato spotted wilt virus. J. Invertebr. Pathol. 33:307–315.

PALIWAL, Y. C. 1980a. Fate of plant viruses in mite vectors. *In* Vectors of Plant Pathogens (ed. K. F. Harris and K. Maramorosch), pp. 357–373. Academic Press, Inc., New York.

PALIWAL, Y. C. 1980b. Relationship of wheat streak mosaic and barley stripe mosaic viruses to vector and nonvector eriophyid mites. Arch. Virol. 63:123–132.

PALIWAL, Y. C., and SINHA, R. C. 1970. On the mechanism of persistence and distribution of barley yellow dwarf virus in an aphid vector. Virology 42:668–680.

PALIWAL, Y. C., and SLYKHUIS, J. T. 1967. Localization of wheat streak mosaic virus in the alimentary canal of its vector *Aceria tulipae* Keifer. Virology 32:344–353.

PEDGLEY, D. E. 1982. Windborne Pests and Diseases: Meteorology of Airborne Organisms. John Wiley & Sons, Inc., New York.

PETERS, D. 1973. Persistent aphid-borne viruses. *In* Viruses and Invertebrates (ed. A. J. Gibbs), pp. 463–475. North-Holland Publishing Company, Amsterdam.

PETERS, D. 1981. Plant rhabdovirus group. Commonwealth Mycological Institute/Association of Applied Biologists. Descriptions of Plant Viruses, No. 244.

PIENKOWSKI, R. L., and MEDLER, J. T. 1964. Synoptic weather conditions associated with long-range movement of the potato leafhopper, *Empoasca fabae,* into Wisconsin. Ann. Entomol. Soc. Am. 57:588–591.

PIRONE, T. P. 1969. Mechanism of transmission of stylet-borne viruses. *In* Viruses, Vectors, and Vegetation (ed. K. Maramorosch), pp. 199–210. John Wiley & Sons, Inc., New York.

PIRONE, T. P. 1977. Accessory factors in nonpersistent virus transmission. *In* Aphids as Virus Vectors (ed. K. F. Harris and K. Maramorosch), pp. 221–225. Academic Press, Inc., New York.

PIRONE, T. P., and HARRIS, K. F. 1977. Nonpersistent transmission of plant viruses by aphids. Annu. Rev. Phytopathol. 15:55–73.

PIRONE, T. P., and MEGAHED, E. 1966. Aphid transmissibility of some purified viruses and viral RNAs. Virology 30:631–637.

PIRONE, T. P., and THORNBURY, D. W. 1983. Role of virion and helper component in regulating aphid transmission of tobacco etch virus. Phytopathology 73:872–875.

PIRONE, T. P., and THORNBURY, D. W. 1984. The involvement of a helper component in nonpersistent transmission of plant viruses by aphids. Microbiol. Sci. 8:191–193.

PLUMB, R. T. 1977. Aphids and virus control on cereals. Proc. 1977 Br. Crop Prot. Conf. Pests Dis. 3:903–913.

PLUMB, R. T. 1983. Barley yellow dwarf virus—a global problem. *In* Plant Virus Epidemiology (ed. R. T. Plumb and J. M. Thresh), pp. 185–198. Blackwell Scientific Publications Ltd., Oxford.

PLUMB, R. T., and LENNON, E. 1982. Aphid infectivity and the infectivity index. Rothamsted Exp. Stn. Rep. Part 1, 1981, pp. 195–197.

POSNETTE, A. F. 1947. Virus diseases of cacao. I. Cacao viruses 1A, 1B, 1C, and 1D. Ann. Appl. Biol. 34:388–402.

QUIOT, J. B., VERBRUGGHE, M., LABONNE, G., LECLANT, F., and MARROU, J. 1979. Écologie et épidemiologie du Virus de la Mosaïque du concombre dans le Sud-Est de la France. IV. Influence des brise-vent sur la répartition contaminations virales dans une culture protegée. Ann. Phytopathol. 11:307–324.

RACCAH, B. 1986. Nonpersistent viruses: epidemiology and control. Adv. Virus Res. 31:387–429.

RACCAH, B., LOEBENSTEIN, G., and BAR-JOSEPH, M. 1976. Transmission of citrus tristeza virus by the melon aphid. Phytopathology 66:1102–1104.

RACCAH, B., BAR-JOSEPH, M., and LOEBENSTEIN, G. 1978. The role of aphid vectors and variation in virus isolates in the epidemiology of tristeza disease. *In* Plant Disease Epidemiology (ed. P. R. Scott and A. Bainbridge), pp. 221–227. Blackwell Scientific Publications Ltd., Oxford.

RAWORTH, D. A. 1984. Population dynamics of the cabbage aphid, *Brevicoryne brassicae* (Homoptera: Aphididae) at Vancouver, British Columbia, V. Simulation. Can. Entomol. 116:895–911.

ROCHOW, W. F. 1969a. Biological properties of four isolates of barley yellow dwarf virus. Phytopathology 59:1580–1589.

ROCHOW, W. F. 1969b. Specificity in aphid transmission of a circulative plant virus. *In* Viruses, Vectors, and Vegetation (ed. K. Maramorosch), pp. 175–198. John Wiley & Sons, Inc., New York.

ROCHOW, W. F., and PANG, E. 1961. Aphids can acquire strains of barley yellow dwarf virus they do not transmit. Virology 15:382–384.

ROSCIGLIONE, B., CASTELLANO, M. A., MARTELLI, G. P., SAVINO, V., and CANNIZZARO, G. 1983. Mealybug transmission of grapevine virus A. Vitis 22:331–347.

ROSE, D. J. W. 1973. Distances flown by *Cicadulina* spp. (Hem. Cicadellidae) in relation to the distribution of maize streak disease in Rhodesia. Bull. Entomol. Res. 62:497–505.

ROSE, D. J. W. 1978. Epidemiology of maize streak disease. Annu. Rev. Entomol. 23:259–282.

ROSENBERG, L. J., and MAGOR, J. I. 1983. A technique for examining the long-distance spread of plant virus diseases transmitted by the brown planthopper *Nilaparvata lugens* (Homoptera: Delphacidae), and other windborne insect vectors. *In* Plant Virus Epidemiology (ed. R. T. Plumb and J. M. Thresh), pp. 229–238. Blackwell Scientific Publications Ltd., Oxford.

ROVAINEN, O. 1973. Viruses and Coccoidea. *In* Viruses and Invertebrates (ed. A. J. Gibbs), pp. 455–462. North-Holland Publishing Company, Amsterdam.

ROVAINEN, O. 1976. Transmission of cocoa viruses by mealybugs (Homoptera: Pseudococcidae). J. Sci. Agric. Soc. Finl. 48:203–304.

ROVAINEN, O. 1980. Mealybugs. *In* Vectors of Plant Pathogens (ed. K. F. Harris and K. Maramorosch), pp. 15–38. Academic Press, Inc., New York.

RUESINK, W. G., and IRWIN, M. E. 1986. Soybean mosaic virus: a model and implications. *In* Plant Virus Epidemics: Monitoring, Modeling and Predicting Outbreaks (ed. G. D. McLean, R. G. Garrett, and W. G. Ruesink), pp. 295–313. Academic Press, Inc., New York.

RUSSELL, G. E. 1970. Beet yellows virus. Commonwealth Mycological Institute/Association of Applied Biologists. Descriptions of Plant Viruses, No. 13.

SAKIMURA, K. 1963. *Frankliniella fusca,* an additional vector for tomato spotted wilt virus, with notes on *Thrips tabaci,* another vector. Phytopathology 53:412–415.

SCOTT, H. A., and FULTON, J. P. 1978. Comparison of the relationship of southern bean mosaic virus and the cowpea strain of tobacco mosaic virus with the bean leaf beetle. Virology 84:197–199.

SELLERS, R. F. 1980. Weather, host and vector—their interplay in the spread of insect-borne animal virus diseases. J. Hyg. Camb. 85:65–102.

SELLERS, R. F., GIBBS, E. P. J., HERNIMAN, K. A. J., PEDGLEY, D. E., and TUCKER, M. R. 1979. Possible origin of the bluetongue epidemic in Cyprus, August 1977. J. Hyg. Camb. 83:547–555.

SELMAN, B. J. 1973. Beetles—phytophagous Coleoptera. *In* Viruses and Invertebrates (ed. A. J. Gibbs), pp. 157–177. Elsevier/North-Holland, Amsterdam.

SHANKS, C. H., Jr. 1965. Seasonal populations of the strawberry aphid and transmission of strawberry viruses in relation to virus control in western Washington. J. Econ. Entomol. 58:316–322.

SHEPHERD, R. J. 1977. Intrinsic properties and taxonomy of aphid-borne viruses. *In* Aphids as Virus Vectors (ed. K. F. Harris and K. Maramorosch), pp. 121–136. Academic Press, Inc., New York.

SHEPHERD, R. J., and HILLS, F. J. 1970. Dispersal of beet yellows and beet mosaic viruses in the inland valleys of California. Phytopathology 60:798–804.

SHIKATA, E. 1981. Reoviruses. *In* Handbook of Plant Virus Infections (ed. E. Kurstak), pp. 423–451. Elsevier/North-Holland, Amsterdam.

SHIVANATHAN, P. 1983. The epidemiology of three diseases caused by whitefly-borne pathogens. *In* Plant Virus Epidemiology (ed. J. T. Plumb and J. M. Thresh), pp. 323–330. Blackwell Scientific Publications Ltd., Oxford.

SHIYOMI, M. 1974. A model of plant-to-plant movement of aphids. III. Studies of actual movement and apparent movement by simulation technique. Res. Popul. Ecol. (Kyoto) 15:148–162.

SIMONS, J. N. 1957. Effect of insecticides and physical barriers on field spread of pepper veinbanding mosaic virus. Phytopathology 47:139–145.

SIMONS, J. N. 1959. Variation in efficiency of aphid transmission of southern cucumber mosaic virus and potato virus Y in pepper. Virology 9:612–623.

SIMONS, J. N. 1960. Factors affecting field spread of potato virus Y in south Florida. Phytopathology 50:424–428.

SIMONS, J. N. 1982. Use of oil sprays and reflective surfaces for control of insect transmitted plant viruses. *In* Pathogens, Vectors, and Plant Diseases (ed. K. F. Harris and K. Maramorosch), pp. 71–93. Academic Press, Inc., New York.

SINHA, R. C. 1973. Viruses and leafhoppers. *In* Viruses and Invertebrates (ed. A. J. Gibbs), pp. 493–511. North-Holland Publishing Company, Amsterdam.

SLYKHUIS, J. T. 1955. *Aceria tulipae* Keifer (Acarina: Eriophyidae) in relation to the spread of wheat streak mosaic. Phytopathology 45:116–128.

SLYKHUIS, J. T. 1969. Mites as vectors of plant viruses. *In* Viruses, Vectors, and Vegetation (ed. K. Maramorosch), pp. 121–141. John Wiley & Sons, Inc., New York.

SLYKHUIS, J. T. 1980. Mites. *In* Vectors of Plant Pathogens (ed. K. F. Harris and K. Maramorosch), pp. 325–336. Academic Press, Inc., New York.

SMITH, K. M. 1965. Plant virus–vector relationships. Adv. Virus Res. 11:61–95.

SOHI, S. S., and SWENSON, K. G. 1964. Pea aphid biotypes differing in bean yellow mosaic virus transmission. Entomol. Exp. Appl. 7:9–14.

STOREY, H. H. 1939. Transmission of plant viruses by insects. Bot. Rev. 5:240–272.

SWENSON, K. G. 1968. Role of aphids in the ecology of plant viruses. Annu. Rev. Phytopathol. 6:351–374.

SYLVESTER, E. S. 1954. Aphid transmission of nonpersistent plant viruses with special reference to the *Brassica nigra* virus. Hilgardia 23:53–98.

SYLVESTER, E. S. 1956. Beet yellows virus transmission by the green peach aphid. J. Econ. Entomol. 49:789–800.

SYLVESTER, E. S. 1969. Virus transmission by aphids—a viewpont. *In* Viruses, Vectors, and Vegetation (ed. K. Maramorosch), pp. 159–173. John Wiley & Sons, Inc., New York.

SYLVESTER, E. S. 1967. Retention of inoculativity in the transmission of pea enation mosaic virus by pea aphids as associated with virus isolates, aphid reproduction and excretion. Virology 32:524–531.

SYLVESTER, E. S. 1980. Circulative and propagative virus transmission by aphids. Annu. Rev. Entomol. 25:257–286.

SYLVESTER, E. S., and SIMONS, J. N. 1951. Relation of plant species inoculated to efficiency of aphids in the transmission of *Brassica nigra* virus. Phytopathology 41:908-910.

TAKAHASHI, Y., and ORLOB, G. B. 1969. Distribution of wheat streak mosaic virus-like particles in *Aceria tulipae.* Virology 38:230–240.

TAKAMI, G., and SMITH, R. F. 1972. Influence of wind and migrant aphid source on the flight and infestation patterns of the spotted alfalfa aphid. Ann. Entomol. Soc. Am. 65:1131–1143.

TAMADA, T., and HARRISON, B. D. 1981. Quantitative studies on the uptake and retention of potato leafroll virus by aphids in laboratory and field conditions. Ann. Appl. Biol. 98:261–276.

TAYLOR, L. R. 1977. Migration and the spatial dynamics of an aphid, *Myzus persicae.* J. Anim. Ecol. 46:411–423.

TAYLOR, L. R. 1984. Assessing and interpreting the spatial distributions of insect populations. Annu. Rev. Entomol. 29:321–357.

TAYLOR, L. R. 1986. The distribution of virus disease and the migrant vector aphid. *In* Plant Virus Epidemics: Monitoring, Modelling and Predicting Outbreaks (ed. G. D. McLean, R. G. Garrett, and W. G. Ruesink), pp. 35–55. Academic Press, Inc., New York.

TAYLOR, L. R., and TAYLOR, R. A. J. 1977. Aggregation, migration and population dynamics. Nature 265:415–421.

TAYLOR, R. A. J. 1985. Migratory behavior in the Auchenorrhyncha. *In* The Leafhoppers and Planthoppers (ed. L. R. Nault and J. G. Rodriquez), pp. 259–288. John Wiley & Sons, Inc., New York.

THRESH, J. M. 1958. The spread of virus disease in cacao. West Afr. Cocoa Res. Inst. Tech Bull. 4.

THRESH, J. M. 1974a. Vector relationships and the development of epidemics: the epidemiology of plant viruses. Phytopathology 64:1050–1056.

THRESH, J. M. 1974b. Temporal patterns of virus spread. Annu. Rev. Phytopathol. 12:111–128.

THRESH, J. M. 1976. Gradients of plant virus diseases. Ann. Appl. Biol. 82:381–406.

THRESH, J. M. 1980. The origins and epidemiology of some important plant virus diseases. Appl. Biol. 5:1–65.

THRESH, J. M. (Ed.) 1981. Pests, Pathogens and Vegetation. Pitman Publishing, Inc., Marshfield, Mass.

THRESH, J. M. 1982. Cropping practices and virus spread. Annu. Rev. Phytopathol. 20:193–218.

THRESH, J. M. 1983a. Progress curves of plant virus disease. Adv. Appl. Biol. 8:1–85.

THRESH, J. M. 1983b. The long-range dispersal of plant viruses by arthropod vectors. Philos. Trans. R. Soc. London Ser. B 302:497–528.

TIMIAN, R. G., and ALM, K. 1973. Selective inbreeding of *Macrosteles fascifrons* for increased efficiency in virus transmission. Phytopathology 63:109–112.

VANDERPLANK, J. E. 1963. Plant Diseases: Epidemics and Control. Academic Press, Inc., New York.

VANDERVEKEN, J. J. 1977. Oils and other inhibitors of nonpersistent virus transmission. *In* Aphids as Virus Vectors (ed. K. F. Harris and K. Maramorosch), pp. 435–454. Academic Press, Inc., New York.

WALLACE, H. R. 1978. Dispersal in time and space: soil pathogens. *In* Plant Disease: An Advanced Treatise, Vol. 2 (ed. J. G. Horsfall and E. B. Cowling), pp. 181–202. Academic Press, Inc., New York.

WALLIN, J. R., and LOONAN, D. V. 1971. Low-level jet winds, aphid vectors, local weather and barley yellow dwarf virus outbreaks. Phytopathology 61:1068–1070.

WALLIN, J. R., PETERS, D., and JOHNSON, L. C. 1967. Low-level jet winds, early cereal aphid and barley yellow dwarf detection in Iowa. Plant Dis. Rep. 51:527–530.

WALTERS, H. J. 1969. Beetle transmission of plant viruses. Adv. Virus Res. 15:339–363.

WALTERS, H. J., LEE, F. N., and JACKSON, K. E. 1972. Overwintering of bean pod mottle virus in bean leaf beetles. Phytopathology (Abstr.) 62:808.

WATSON, M. A. 1936. Factors affecting the amount of infection obtained by aphis transmission of the virus Hy III. Philos. Trans. R. Soc. London Ser. B 226:457–489.

WATSON, M. A. 1946. The transmission of beet mosaic and beet yellows viruses by aphids; a comparative study of a nonpersistent and a persistent virus having host plants and vectors in common. Proc. R. Soc. London Ser. B 133:200–219.

WATSON, M. A. 1967. The epidemiology of aphid-transmistted plant-virus diseases. Outlook Agric. 6:155–166.

WATSON, M. A. 1972. Transmission of plant viruses by aphids. *In* Principles and Techniques in Plant Virology (ed. C. I. Kado and H. O. Agrawal), pp. 131–167. Van Nostrand Rheinhold Company, Inc., New York.

WATSON, M. A., and HEALY, M. J. R. 1953. The spread of beet yellows and beet mosaic viruses in the sugar beet crops. II. The effect of aphid numbers on disease incidence. Ann. Appl. Biol. 40:38–59.

WATSON, M. A., and PLUMB, R. T. 1972. Transmission of plant pathogenic viruses by aphids. Annu. Rev. Entomol. 17:425–452.

WATSON, M. A., and ROBERTS, F. M. 1939. A comparative study of the transmission of *Hyoscyamus* virus 3, potato virus Y and cucumber virus 1 by the vectors *Myzus persicae* (Sulz.), *M. circumflexus* (Buckton), and *Macrosiphum qei* (Koch). Proc. R. Soc. London Ser. B 127:543–576.

WATSON, M. A., and ROBERTS, F. M. 1940. Evidence against the hypothesis that certain plant viruses are transmitted mechanically by aphids. Ann. Appl. Biol. 27:227–233.

WATSON, M. A., HEATHCOTE, G. D., LAUKNER, F. B., and SOWREY, P. A. 1975. The use of weather data and counts of aphids in the field to predict the incidence of yellowing viruses of sugar-beet crops in England in relation to the use of insecticides. Ann. Appl. Biol. 81:181–198.

WAY, M. J., CAMMEL, M. E., TAYLOR, L. R., and WOIWOD, I. P. 1981. The use of egg counts and suction trap examples to forecast the infestation of spring-sown field beans, *Vicia faba,* by the black bean aphid, *Aphis fabae.* Ann. Appl. Biol. 98:21–34.

WIKTELIUS, S. 1977. The importance of southerly winds and other weather data on the incidence of sugar beet yellowing viruses in southern Sweden. Swed. J. Agric. Res. 7:89–95.

WOOLSTON, C. J., COVEY, S. N., PENSWICK, J. R., and DAVIES, J. W. 1983. Aphid transmission and a polypeptide are specified by a defined region of the cauliflower mosaic virus genome. Gene 23:15–23.

ZADOKS, J. C., and SCHEIN, R. D. 1979. Epidemiology and Plant Disease Management. Oxford University Press, Oxford.

ZETTLER, F. W., and WILKINSON, R. E. 1966. Effect of probing behavior and starvation of *Myzus persicae* on transmission of bean common mosaic virus. Phytopathology 56:1079–1082.

ZEYEN, R. J., STROMBERG, E. L., and KUEHNAST, E. 1987. Long-range aphid transport hypothesis for maize dwarf mosaic: history and distribution in Minnesota, U.S.A. Ann. Appl. Biol. 111:325–336.

ZINK, F. W., GROGAN, R. G., and WELCH, J. E. 1956. The effect of the percentage of seed transmission upon subsequent spread of lettuce mosaic virus. Phytopathology 46:662–664.

ZITTER, T. A. 1977. Epidemiology of aphid-borne viruses. *In* Aphids as Virus Vectors (ed. K. F. Harris and K. Maramorosch), pp. 385–412. Academic Press, Inc., New York.

4

World Distribution of Soilborne Mycoparasites

An Evaluation

P. B. Adams

INTRODUCTION

There are a number of reasons for attempting to learn the geographical distribution of soilborne fungi. For plant pathogens the reasons are obvious: One wants to know if a pathogen of an economic crop is in the region of the crop production area. Conversely, if it is known that a particular area is free of a specific pathogen, that area may be chosen for growth of a particular crop. In some parts of the United States, the value of land is negatively correlated with the degree of infestation by a particular soilborne plant pathogen.

The geographical distribution of soilborne beneficial fungi, such as mycoparasities, should also be determined for several reasons. Mycoparasites and other beneficial fungi are now being studied intensively to determine their potential as microbial pesticides (Ayers and Adams, 1981a). Knowledge of the distribution of a mycoparasite within a state or province, within a country, and throughout the world would aid research. Such information is useful for locating areas for research experiments. It also facilitates the shipment of specimens across state or national boundaries and may stimulate research on a

Note: Research Plant Pathologist, Biocontrol of Plant Diseases Laboratory, Plant Sciences Institute, U.S. Department of Agriculture, Beltsville, MD 20705, US.

particular mycoparasite by other scientists. To obtain an experimental use permit or registration of a microbial pesticide the U.S. Environmental Protection Agency (EPA) requires information on the geographical distribution of the beneficial organisms (Moore, 1984). Additional benefits from research devoted to the estimation of the geographical distribution of beneficial fungi are that it can provide information on the natural populations of these organisms, as well as being a source of additional isolates with which to study.

Soil Mycoparasites

There are relatively few well-documented cases of soilborne fungi that are known to be mycoparasitic on plant pathogenic fungi in their natural habitats. There are, however, four such fungi whose geographical distributions have been estimated: *Coniothyrium minitans* Campbell; *Sporidesmium sclerotivorum* Uecker, Ayers, & Adams; *Teratosperma oligocladum* Uecker, Ayers, & Adams; and *Laterispora brevirama* Uecker, Ayers, & Adams. This chapter is restricted to these four mycoparasites, but the discussion could apply to most soilborne fungi.

Coniothyrium minitans was first described by Campbell (1947) and was shown to be a mycoparasite of sclerotia of *Sclerotinia sclerotiorum* formed on a host plant. It was subsequently shown to parasitize sclerotia of other *Sclerotinia* species, *Botrytis* species, and *Sclerotium cepivorum* and to cause their decay in natural soil (Ayers and Adams, 1981a). Huang (1977) showed that this fungus was involved in the natural destruction of sclerotia of *S. sclerotiorum* in Canadian sunflower fields. *C. minitans* invades host sclerotia in soil or on host plants, colonizes the sclerotia, and produces numerous pycnidia in and on the sclerotia.

No one has attempted to determine the geographical distribution of *C. minitans.* The distribution of this fungus is known only by persons discovering it on sclerotia in regions of their own country. To date, *C. minitans* has been reported in Australia (Merriman et al., 1979), Canada (Huang and Hoes, 1976), England (Tribe, 1957), Finland (Ervio, 1965), Germany (Schmidt, 1970), Hungary (Voros, 1969), New Zealand (Jarvis and Hawthorne, 1971), Poland (Zub, 1960), Scotland (Jones and Watson, 1969), United States (Campbell, 1947), and the Soviet Union (Fedulova, 1983).

Sporidesmium sclerotivorum was first described in 1978 (Uecker et al., 1978) on sclerotia of *Sclerotinia minor* from a field in Beltsville, Maryland. This dematiaceous hyphomycete was shown to infect sclerotia of *Sclerotinia* species, *Botrytis cinerea,* and *Sclerotium cepivorum* (Ayers and Adams, 1981a) and to cause their destruction in natural soil. It was also shown to be involved in natural biological control of onion white rot caused by *S. cepivorum* in New Jersey and *Sclerotinia* stem rot of potato caused by *S. sclerotiorum* in Oregon (Adams and Ayers, 1981). This fungus parasitizes sclerotia of host fungi and then grows out several centimeters into the surrounding soil mass to infect healthy sclerotia. In the process, it produces thousands of new macroconidia (Adams et al., 1984).

Teratosperma oligocladum, first described in 1980 (Uecker et al., 1980), is another unusual hyphomycete with a known host range similar to that of *S. sclerotivorum* (Adams and Ayers, 1985). This mycoparasite behaves in soil in a manner very similar to *S. sclerotivorum* (Ayers and Adams, 1981b).

Laterispora brevirama is an even more unusual fungus, first described in 1982 (Uecker et al., 1982). This fungus was first thought to be a mycoparasite of sclerotia of *Sclerotinia* species. Further research showed that it was actually a mycoparasite of *S. sclerotivorum* and *T. oligocladum,* which were parasitizing sclerotia of *Sclerotinia* species (Ayers and Adams, 1985). Thus *L. brevirama* is a mycoparasite of two mycoparasites of at least five plant pathogens.

All of the known work on the geographical distribution of the latter three mycoparasites has been done in two laboratories: those of Adams and Ayers and of Coley-Smith. *Sporidesmium sclerotivorum* was detected in soil samples from 13 of 22 states in the United States, four provinces of Canada, and in Australia, England (J. R. Coley-Smith, personal communication), Finland, Japan, and Norway (Adams and Ayers, 1981, 1985). *Teratosperma oligocladum* was detected in England (Parfitt et al., 1983), Japan, and the United States (Adams and Ayers, 1985; Uecker et al., 1980). *Laterispora brevirama* is known to exist in Australia, Canada, England, Finland, Japan, and the United States (Adams and Ayers, 1985; Parfitt et al., 1984; Uecker et al., 1982).

WORLD DISTRIBUTION

Table 1 lists the countries in which these four mycoparasites are known to exist. However, the world distribution of these fungi is surely far greater than that indicated in Table 1; large portions of the world such as Central and South America, the Soviet Union, Africa, and Southeast Asia have yet to be sampled to any extent. Even countries for which a mycoparasite has not been detected (Table 1) does not mean that it does not have indigenous populations of the fungus. For many of those countries only a limited number of soil samples were assayed. For all countries except the United Kingdom and the United States only 11 soil samples or less were assayed for the presence of *S. sclerotivorum, L. brevirama,* and *T. oligocladum* (Adams and Ayers, 1985). Usually, only two to four samples were assayed from each country. Recognizing these limitations, it is somewhat surprising and revealing that these fungi were detected in as many countries as indicated in Table 1.

Coniothyrium minitans appears to be the most widely distributed mycoparasite. *Sporidesmium sclerotivorum* and *L. brevirama* are both widely distributed throughout the world. All three fungi appear in both the northern and southern hemispheres. *Teratosperma oligocladum* appears to be much more restricted in its distribution than the other three mycoparasites. Even though

TABLE 1 Known Geographical Distribution of Four Soilborne Mycoparasites[a]

Country	*Coniothyrium minitans*	*Sporidesmium sclerotivorum*	*Laterispora brevirama*	*Teratosperma oligocladum*
Australia	+	+	+	−
Bermuda	?	−	−	−
Brazil	?	−	−	−
Canada	+	+	+	−
Chile	?	−	−	−
China	?	−	−	−
Egypt	?	−	−	−
Finland	+	+	+	−
France	?	−	−	−
Germany	+	−	−	−
Hungary	+	?	?	?
Japan	?	+	+	+
Lebanon	?	−	−	−
Netherlands	?	−	−	−
New Zealand	+	?	?	?
Norway	?	+	−	−
Pakistan	?	−	−	−
Poland	+	?	?	?
South Africa	?	−	−	−
Soviet Union	+	?	?	?
United Kingdom	+	+	+	+
United States	+	+	+	+

[a]+, Detected; −, not detected; ?, not determined.

it was detected only in the widely separated countries of Japan, United Kingdom, and United States, *T. oligocladum* is probably more widly distributed than is reported.

The geographical location of the center of origin of an organism is always interesting speculation. In the case of these mycoparasites, it is also a hazardous speculation because of the limited number of observations. The origin of *C. minitans* is very difficult to determine because there are no negative observations on which to base one's estimate, but it could very well be Europe. Similarly, the origin of *T. oligocladum* is difficult to assess because it is known to exist in three widely separated geographical areas. Although *S. sclerotivorum* is widely distributed in the world, it is very common in Canada and the United States and one could speculate that it originated in North America. Since *L. brevirama* is a mycoparasite of *S. sclerotivorum* and *T. oligocladum,* it might have originated with one of these mycoparasites.

Sporidesmium sclerotivorum has probably been the most intensively studied mycoparasite in terms of geographical distribution. In the United States, this mycoparasite has been detected in 13 of 22 states and in 48 of 122 (39%) samples (Table 2). It has also been detected in 7 of 17 countries and in 64 of 329 (19%) soil samples from around the world (Table 3).

TABLE 2 Distribution of *Sporidesmium sclerotivorum* in the United States as Determined by Baiting Soil Samples with Sclerotia of *Sclerotinia minor*

State	Number of Soil Samples	Number containing *S. sclerotivorum*	Percent containing *S. sclerotivorum*
Arizona	1	1	100
California	13	5	38
Connecticut	1	0	0
Florida	5	0	0
Idaho	4	0	0
Illinois	2	0	0
Louisiana	1	1	100
Maryland	18	9	50
Michigan	1	1	100
Nebraska	5	0	0
New Jersey	21	13	62
New York	9	4	44
North Carolina	1	1	100
North Dakota	1	0	0
Ohio	10	2	20
Oklahoma	1	0	0
Oregon	5	4	80
Texas	4	0	0
Vermont	1	0	0
Virginia	5	3	60
Washington	7	1	14
Wisconsin	6	3	50

METHODOLOGIES FOR DETERMINING DISTRIBUTION

The limitations of the methods used in estimating geographical distribution must be understood and taken into consideration in interpreting the results. For example, the nature of the fungus (fungi) in question, method and sensitivity of the fungal assay methods, and the method of selecting and obtaining the soil sample will all affect the recovery of the fungi.

Sampling within Fields

The nature of the fungus determines where it will be located in a geographical region and where within a field one is apt to detect the fungus. The mycoparasites described above are, in nature, obligate parasites of sclerotia of a limited number of plant pathogens. Thus if the host fungi have never been in a particular field, it is unlikely that the mycoparasite will be detected in that field. Some fungi appear to be restricted to a particular horizon in the soil profile. For example, *Rhizoctonia solani* is restricted to the top 10 cm of the soil profile (Papavizas et al., 1975). Thus for this fungus one should restrict

TABLE 3 World Distribution of *Sporidesmium sclerotivorum* as Determined by Baiting Soil Samples with Sclerotia of *Sclerotinia minor*

Country	Number of Fields Sampled	Number of Fields with *S. sclerotivorum*	Percent of Fields with *S. sclerotivorum*
Australia	10	4	40
Bermuda	3	0	0
Canada	11	8	73
Chile	2	0	0
China	2	0	0
Egypt	4	0	0
Finland	4	1	25
France	5	0	0
Germany	3	0	0
Japan	3	1	33
Lebanon	2	0	0
Netherlands	4	0	0
Norway	3	1	33
Pakistan	3	0	0
South Africa	3	0	0
United Kingdom	145	1	1
United States	122	48	39

soil sampling to this depth. It is not known whether such a relationship exists with mycoparasites. Except for *C. minitans,* soil samples for the mycoparasites were requested from various persons in various countries from fields with a history of diseases caused by species of *Sclerotinia* or *Sclerotium cepivorum,* host fungi for *S. sclerotivorum* and *T. oligocladum.*

The method of collecting a soil sample from a particular field will affect the probability of detecting the fungus in question. No soilborne fungus is uniformly distributed in a field. The distribution of sclerotia of *S. minor,* a host for the four mycoparasites mentioned above, and other other plant pathogens in soil has often been described by the negative binomial distribution (Dillard and Grogan, 1985). Under some assumptions this means that the sclerotia of the plant pathogen are in clumps or clusters in the field in a random pattern (see Chapter 5).

If the hosts of a mycoparasite are located in a field in a spatial pattern that follows negative binomial distribution, it is assumed that the spatial pattern of the mycoparasite also follows a negative binomial distribution. To overcome this situation, a number of subsamples should be taken from a field and the subsamples pooled and thoroughly mixed to obtain a composite soil sample that approximates that of the entire field.

An example of the nonuniform and also nonrandom pattern of a sclerotial plant pathogen in a small portion of a field is that reported for *S. cepivorum,* a host of three of the mycoparasites (Adams, 1981). A small portion (6×122 m) of one *S. cepivorum*-infested field was divided into 16 plots

(1.5 × 30.5 m). From each plot 10 subsamples of soil to a depth of 15 cm were taken and assayed for the number of sclerotia. The inoculum density of the plant pathogen varied from 6 to 136, with an average of 40.2 sclerotia per 100 g of soil (Fig. 1). On a larger scale, a 2.4-ha *S. cepivorum*-infested field was divided into thirds and the inoculum density determined from a soil sample consisting of 50 subsamples. The inoculum densities of the three 0.8-ha portions were 0, 6, and 12 sclerotia per 100 g of soil (Fig. 2A). When this 2.4-ha field was further subdivided into 0.2-ha portions and sampled (50 subsamples per sample), the inoculum densities ranged from 0 to 36 sclerotia per 100 g of soil, with an average of 7.5 sclerotia per 100 g of soil (Fig. 2B). The same study (Adams, 1981) revealed that it was necessary to take 50 subsamples for each 0.2-ha portion of a field to obtain a soil sample that reasonably represented that 0.2-ha field. In addition, the 50 subsamples should be taken in a uniform pattern as shown in Fig. 3.

It is not known how the soil samples were selected and taken for the work on *C. minitans*. For the other three mycoparasites, cooperators were asked simply to collect soil samples from fields with a history of diseases caused by host plant pathogens and to send the samples to Beltsville, Maryland, to be assayed. It is not known whether the soil sample came from a single location in the field or whether the sample represented many subsamples.

Processing of Samples

Some fungi, including *S. sclerotivorum,* cannot survive in air-dried soil. Thus how the soil sample is handled after the sample is taken should be considered, which requires some knowledge of the test fungus. For the samples collected to detect *L. brevirama, S. sclerotivorum,* and *T. oligocladum,* the cooperators were requested to collect and put the soil samples in a plastic bag to prevent the sample from drying during shipment to Beltsville, Maryland.

The method of detecting the mycoparasite in the soil sample and its sensitivity will have a marked effect on the results obtained. For the mycoparasites

84	8	6	49
19	8	50	136
10	48	81	23
40	62	11	8

Figure 1 Field diagram illustrating the variation in the inoculum density of *Sclerotium cepivorum* in a small area of a field. Each number represents the number of sclerotia per 100 g of soil for the plot. Each plot was 1.5 × 30.5 m.

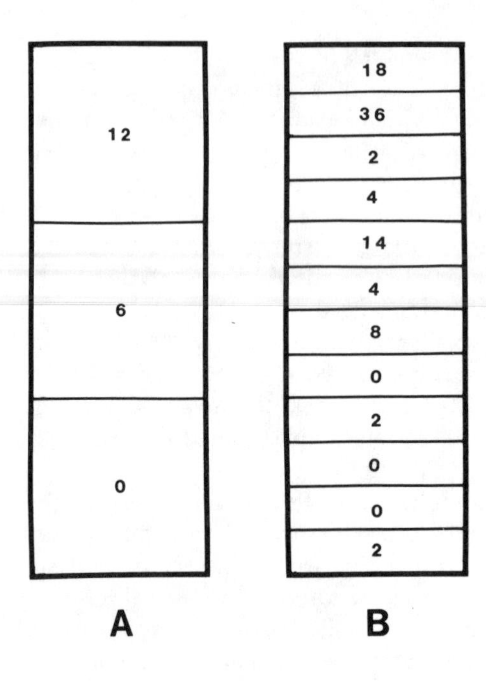

Figure 2 Field diagram illustrating the variation in the inoculum density of *Sclerotium cepivorum* in a 2.4-ha field. A, field divided into 0.8-ha portions. B, the same field divided into 0.2-ha portions. Each number represents the number of sclerotia per 100 g of soil for that portion of the field.

in question, no dilution-plate method is available. *C. minitans* has been detected by plating out either natural or introduced sclerotia of a host fungus on a substrate of moist filter paper, moist sand, or on an agar medium. The other three mycoparasites were all detected by plating out sclerotia of *S. minor* introduced into the soil sample 4 or more weeks previously. Thus the mycoparasites were all detected by various baiting methods using sclerotia of a host fungus. The sensitivities of these methods is not well known but in the case of *S. sclerotivorum* it is estimated to be about 10 macroconidia per gram of soil (Adams and Ayers, 1981). If the inoculum densities in Figs. 1 and 2 were those of *S. sclerotivorum* instead of *S. cepivorum,* the mycoparasite would not have been detected in 4 of the 16 plots in Fig. 1 and 9 of the 12 plots in Fig. 2B.

With the limitations and problems associated with collecting and assaying soil samples for soil fungi, especially mycoparasites, as outlined above it is surprising that the four mycoparasites were detected as frequently as they

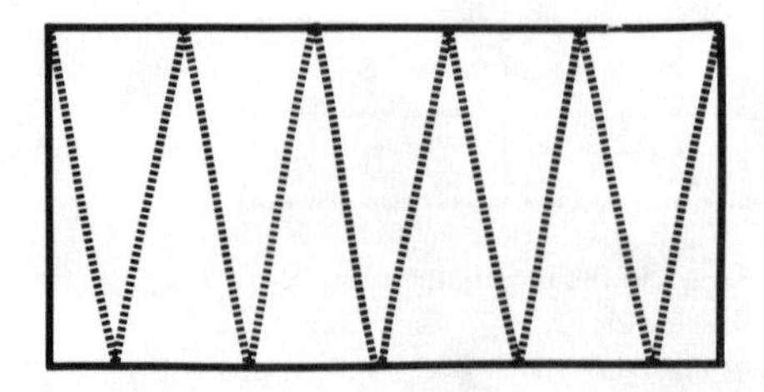

Figure 3 Path traveled across a 0.2-ha field to collect subsamples (dotted line for the assessment of *Sclerotium cepivorum*). On each traverse of the field five subsamples were taken for a total of 50 subsamples per field.

were. This is especially true when in some cases only two or three soil samples from a particular country were assayed.

MECHANISMS OF DISPERSAL

Regardless of where each fungus originated, it has subsequently spread around the world. These soilborne fungi are not expected to have been dispersed long distances by wind or water. Because of the similarity of isolates of a species from widely separated geographical regions, they most likely were spread in a form very similar to their present characteristics. That is, in nature they are obligate parasites of sclerotia of a number of soilborne plant pathogens. Thus they were most likely spread in soil or on their host sclerotia. If this is true, they were probably spread around the world by human activities. From 1500 to about 1900, Europeans were exploring and colonizing new territories throughout the world. During this period, potted plants containing soil were brought from Europe to these new territories and exotic potted plants made the return voyage. In this manner, numerous plant pathogens must have been transported between the Old World and the New World, and vice versa. Many of the mycoparasites of these plant pathogens must have made the same voyage.

Three of these mycoparasites are pathogens of sclerotia of *Sclerotinia* species. Even today the sclerotia of these plant pathogens travel around the world in ships containing seed of various host crops, such as sunflower, soybean, peanuts, alfalfa, and other agricultural commodities. *C. minitans* has been shown to infect sclerotia of *S. sclerotiorum* in the stocks of sunflower. Sclerotia are known to be present in international shipments of soybean, sunflower, and rapeseed. *Sclerotinia* species have a very wide host range, causing disease on at least 361 plant species in 64 plant families (Purdy, 1979). Many of these plants are common weeds in agricultural crops. It has been shown that *S. sclerotiorum* can parasitize a weed in a soybean field and the sclerotia produced on the weed can be found in the harvested soybeans (Adams et al., 1983). Thus sclerotia of *Sclerotinia* species, and presumably their mycoparasites, are still being exported from the United States, Canada, and probably other countries along with such commodities as alfalfa, soybean, rape, and sunflower seeds.

International commerce is probably not the only method for the continuing dispersal of soilborne fungi. Baker (1966) found 65 species of fungi on the footware of passengers on planes arriving at Honolulu International Airport. A number of these species were plant pathogens, of which a number were not known to exist in Hawaii. Zadoks (1967) questioned whether traveling scientists are involved in the international dispersal of fungi.

Tourists are not the only travelers who often escape detection by plant quarantine officials. Suthers (1985) showed that North American migratory thrushes, finches, sparrows, and warblers could transport cellular slime molds (Dictyostelids) from eastern North America to Central America. Presumably, these birds and possibly other migratory birds could transport soilborne plant pathogenic fungi and their mycoparasites between summer and winter feeding grounds.

REFERENCES

ADAMS, P. B. 1981. Forecasting onion white rot diseases. Phytopathology 71:1178–1181.

ADAMS, P. B., and AYERS, W. A. 1981. *Sporidesmium sclerotivorum:* distribution and function in natural biological control of sclerotial fungi. Phytopathology 71:90–93.

ADAMS, P. B., and AYERS, W. A. 1985. The world distribution of the mycoparasites *Sporidesmium sclerotivorum, Teratosperma oligocladum* and *Laterispora brevirama.* Soil Biol. Biochem. 17:583–584.

ADAMS, P. B., MAROSE, B. H., and DUTKY, E. M. 1983. Cocklebur: a new host for several *Sclerotinia* species. Plant Dis. 67:484–485.

ADAMS, P. B., MAROIS, J. J., and AYERS, W. A. 1984. Population dynamics of the mycoparasite, *Sporidesmium sclerotivorum,* and its host, *Sclerotinia minor,* in soil. Soil Biol. Biochem. 16:627–633.

AYERS, W. A., and ADAMS, P. B. 1981a. Mycoparasitism and its application to biological control of plant diseases. *In* Biological Control in Crop Production (ed. G. C. Papavizas), Chap. 6. Beltsville Symposia in Agricultural Research 5:91–103. Allanheld, Osmun & Co. Publishers, Inc., Totowa, N.J.

AYERS, W. A., and ADAMS, P. B. 1981b. Mycoparasitism of sclerotial fungi by *Teratosperma oligocladum.* Can. J. Microbiol. 27:886–892.

AYERS, W. A., and ADAMS, P. B. 1985. Interaction of *Laterispora brevirama* and the mycoparasites *Sporidesmium sclerotivorum* and *Teratosperam oligocladum.* Can. J. Microbiol. 31:786–792.

BAKER, G. E. 1966. Inadvertent distribution of fungi. Can. J. Microbiol. 12:109–112.

CAMPBELL, W. A. 1947. A new species of *Coniothyrium* parasitic on sclerotia. Mycologia 39:190–195.

DILLARD, H. R., and GROGAN, R. G. 1985. Relationship between sclerotial pattern and density of *Sclerotinia minor* and the incidence of lettuce drop. Phytopathology 75:90–94.

ERVIO, L. R. 1965. Certain parasites of fungal sclerotia. J. Sci. Agric. Soc. Finl. 37:1–6.

FEDULOVA, T. Yu. 1983. Activity of the fungus *Coniothyrium minitans* hyperparasite of soft rot. Sov. Agric. Sci. (Engl. Trans. Dokl, Vses. Ordena Lenina Ordena Trud. Krasnogo Znameni Akad. Skh. Nauk Im. VI Lenina) 0 (10):75–78.

HUANG, H. C. 1977. Importance of *Coniothyrium minitans* in survival of sclerotia of *Sclerotinia sclerotiorum* in wilted sunflower. Can. J. Bot. 55:289–295.

HUANG, H. C., and HOES, J. A. 1976. Penetration and infection of *Sclerotinia sclerotiorum* by *Coniothyrium minitans.* Can. J. Bot. 54:406–410.

JARVIS, W. R., and HAWTHORNE, B. T. 1971. 17th Annu. Rep. 1970, Scott. Hortic. Res. Inst. (Rev. Plant Pathol. 51:11, 1972).

JONES, D., and WATSON, D. 1969. Parasitism and lysis by soil fungi of *Sclerotinia sclerotiorum* (Lib.) de Bary, a phytopathogic fungus. Nature (London) 224:287-288.

MERRIMAN, P. R., PYWELL, M., HARRISON, G., and NANCARROW, 1979. Survival of sclerotia of *Sclerotinia sclerotiorum* and effects of cultivation practices on disease. Soil Biol. Biochem. 11:567-570.

MOORE, J. A. 1984. Microbial pesticides; interim policy on small scale field testing. Fed. Regist. 49(202):40659-40661, October 17, 1984, Notices.

PAPAVIZAS, G. C., ADAMS, P. B., LUMSDEN, R. D., LEWIS, J. A., DOW, R. L., AYERS, W. A., and KANTZES, J. G. 1975. Ecology and epidemiology of *Rhizoctonia solani* in field soil. Phytopathology 65:871-877.

PARFITT, D., COLEY-SMITH, J. R., and JEVES, T. M. 1983. *Teratosperma oligocladum,* a mycoparasite of fungal sclerotia. Plant Pathol. 32:459-460.

PARFITT, D., COLEY-SMITH, J. R., and MCHALE, T. 1984. *Laterispora brevirama* on sclerotia of *Sclerotinia.* Plant Pathol. 33:441-443.

PURDY, L. H. 1979. *Sclerotinia sclerotiorum:* history, diseases and symptomatology, host range, geographical distribution, and impact. Phytopathology 69:875-880.

SCHMIDT, H. H. 1970. Untersuchungen über die Lebensdauer der Sklerotien von *Sclerotinia sclerotiorum* (Lib.) de Bary im Boden unter dem Einfluss verschiedener Pflanzenarten und nach Infektion mit *Coniothyrium minitans* Campb. Arch. Pflanzenschutz 6:321-334.

SUTHERS, H. B. 1985. Ground-feeding migratory songbirds as cellular slime mold distribution vectors. Oecologia (Berlin) 65:526-530.

TRIBE, H. T. 1957. On the parasitism of *Sclerotinia trifoliorum* by *Coniothyrium minitans.* Trans. Br. Mycol. Soc. 40:489-499.

UECKER, F. A., AYERS, W. A., and ADAMS, P. B. 1978. A new hyphomycete on sclerotia of *Sclerotinia sclerotiorum.* Mycotaxon 7:275-282.

UECKER, A., AYERS, W. A., and ADAMS, P. B. 1980. *Teratosperma oligocladum,* a new hyphomycetous mycoparasite on sclerotia of *Sclerotinia sclerotiorum, S. trifoliorum,* and *S. minor.* Mycotaxon 10:421-427.

UECKER, F. A., AYERS, W. A., and ADAMS, P. B. 1982. *Laterispora brevirama,* a new hyphomycete on sclerotia of *Sclerotinia minor.* Mycotaxon 14:491-496.

VOROS, J. 1969. *Coniothyrium minitans* Campbell, a new hyperparasitic fungus in Hungary. Acta Phytopathol. Acad. Sci. Hung. 4:221-227.

ZADOKS, J. C. 1967. International dispersal of fungi. Neth. J. Plant Pathol. 73(Suppl 1):61-80.

ZUB, J. 1960. Nowy dla Polski gatunek grzyba: *Coniothyrium minitans* Campb. Nadpasozyt raka koniczymowego (*Sclerotinia trifoliorum*) Erikss. Biul. Inst. Ochr. Rosl. Poznań 9:171-180 (Rev. Appl. Mycol. 40:612, 1961).

5
Dynamic Nature of Within-Field Disease and Pathogen Distributions

L. V. Madden

INTRODUCTION

A simple but important principle of biology is that organisms are not equal in number at all locations. Similarly, organisms are not equally associated with others of the same population (or species) at all locations. This spatial heterogeneity is the result of environmental and genetic heterogeneity and reproductive population growth acting on the processes of movement (spread), reproduction, and mortality (Taylor, 1984).

Taylor (1984) stated that "spatial distribution is one of the most characteristic ecological properties of species." Heltshe and Ritchey (1984) reflected the belief of many in stating that "knowledge of the spatial distribution of a plant or animal population provides valuable clues as to the factors that are important to its existence in nature." Although some researchers realized the value of assessing the pattern of diseased plants within fields more than 40 years ago (Bald, 1937; Vanderplank, 1946), most interest in this topic developed after about 1980. This book discusses many of the features and implications of spatial heterogeneity of plant pathogens and plant diseases, from the

Note: Department of Plant Pathology, The Ohio State University, Ohio Agricultural Research and Development Center, Wooster, OH 44691, US.

global to the field level. This chapter deals with the spatial variability of pathogens or disease within fields and the dynamic aspects of this variation. In particular, three questions are addressed: (1) what information on disease progress in time or pathogen reproductive strategy is discernible from knowledge of spatial pattern; (2) do patterns change with time in a repeatable, predictable manner; and (3) what is the relationship, if any, between measurements of spatial distribution and other epidemic characteristics such as rate of disease increase? To deal properly with these issues, it is necessary to present the approaches that are commonly used to assess and compare spatial distributions.

SPATIAL DISTRIBUTIONS

There are three more-or-less distinct approaches for assessing spatial distributions of organisms or any other variable (Cliff and Ord, 1981; Upton and Fingleton, 1985). The first approach involves *spatial point patterns.* Here, "points" are discrete, countable, individuals, such as insects, diseased plants, or lesions. Alternatively, "points" can refer to the discrete locations of individuals in a plane (e.g., field). Spatial distribution is determined either by determining the distances between points (e.g., diseased plants) or by counting points within samples. Samples, generally, comprise contiguous quadrats, but random samples from a field are also used.

The second approach involves *spatial autocorrelation analysis.* Essentially a two-dimensional extension of autocorrelation in time series analysis, this approach assesses the patterns of values at specific locations, instead of the patterns of locations as in the first approach. Continuous (e.g., disease severity) as well as discrete (e.g., numbers of diseased plants) data can be analyzed with spatial autocorrelation. With typical epidemiological studies, locations correspond to contiguous quadrats where some measurements of pathogen or disease levels are obtained. For practical purposes, therefore, the same data on disease intensity can be analyzed with spatial point pattern methodology or with spatial autocorrelation.

The final approach can be called *trend surface analysis* and entails the development of models to describe some random variable as a function of absolute or relative location on a surface (Cliff and Ord, 1981). More specifically, independent variables comprise field locations (or distances) and the dependent variable is the variable assessed at each location. Cliff and Ord (1981) call these models "reactive," in contrast to the "interactive" one of spatial autocorrelation. One of the best examples of a trend surface model in epidemiology is Gregory's (1973) gradient model for the spread of plant diseases. Trend surface analysis is not discussed further in this chapter. Minogue (1986; Chapter 6) should be read for many aspects of this approach.

Some researchers have classified the approaches for assessing patterns differently from that presented here. For example, Campbell and Noe (1985) considered distance methods as being separate from spatial point patterns for practical reasons. I group distance methods with point patterns because they are based on the same discrete data theory. Plant pathologists seldom find it convenient or even possible to measure the distances between dieseased plants, although there are some noteworthy exceptions (Gray et al., 1986; Marcus et al., 1984). Therefore, relatively little information on the assessment of spatial distribution using distance methods is presented in this chapter.

Random, Uniform, and Clustered Patterns

One can consider three types of spatial distributions or patterns: (1) regular or uniform; (2) random; and (3) aggregated, clumped, or clustered (Fig. 1). As drawn, there is little question about the pattern of points in Fig. 1a. Unfortunately, often a visual map of the field is not sufficient to make a definitive conclusion about distribution (Fig. 1b). Statistics, then, are necessary to describe and characterize the spatial distribution.

It is not easy to describe adequately what is meant by a random spatial

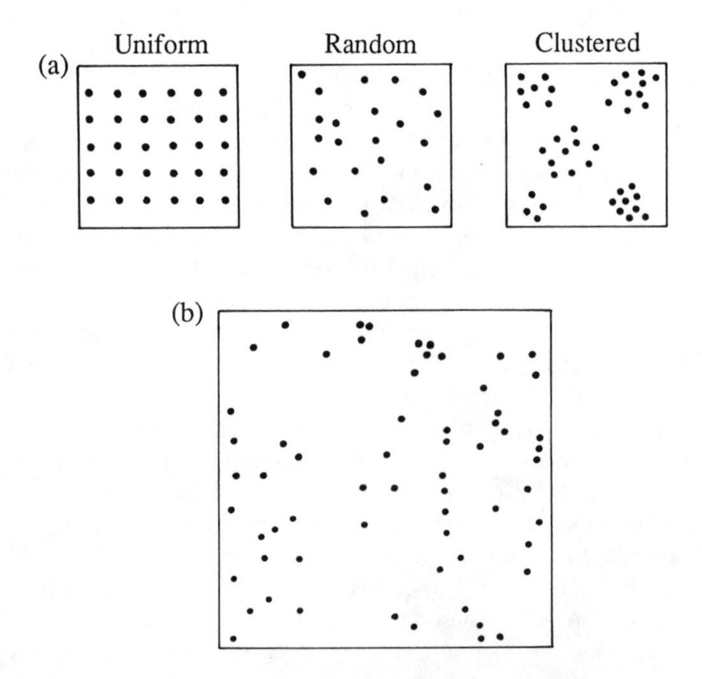

Figure 1 Hypothetical spatial point patterns. Each point can represent, for example, one infected plant: (a) unambiguous uniform, random, and clustered patterns; (b) a pattern for which statistics are needed for evaluation.

pattern of points or other values on a two-dimensional surface. Some of the ways of defining a random pattern, with varying degrees of rigor, include:

1. Every point on a surface has an equal and low probability of being occupied by an individual (e.g., every plant in a field has an equal and low probability of being infected);
2. Knowing the location of any one individual on a surface provides no information as to the location of any other individual;
3. No individual can influence any other in any way;
4. Individuals are independently and randomly *assigned* to available units; and
5. Every pair of units i and j (e.g., locations, quadrats) in the study area have measurements X_i and X_j that are uncorrelated or independent.

(Cliff and Ord, 1981; Pielou, 1977; Upton and Fingleton, 1985).

There are no degrees to randomness—a pattern is or is not random. Nonrandomness, either in the form of uniformity or clustering, is a matter of degree. Statistically, one can consider randomness as the null hypothesis that is tested with actual data. Epidemiologists usually are not interested in uniform patterns because they are not expected to occur except at very high disease levels when spatial heterogeneity is very low. Aggregation of pathogens and diseased plants is expected, however, and is a great concern to many epidemiologists. As with randomness, aggregation can be defined in various ways:

1. Every point on a surface does not have an equal probability of being occupied by an individual (e.g., every plant in a field does not have an equal probability of being infected);
2. Knowing the location of any one individual on a surface provides some information as to the location of the other individuals;
3. Occupancy of a location by an individual increases the probability of other locations being occupied; and
4. Spatial condition in which density (e.g., disease level) is more locally condensed than at random.

(Cliff and Ord, 1981; Pielou, 1977; Upton and Fingleton, 1985).

Other descriptions are possible, but these listed here should be sufficient for most purposes. An example of a field with a clustered pattern of infected plants is given in Fig. 2. (The statistics that describe the clustered pattern are discussed in later sections.) The quantification of aggregation is done in different ways, depending on whether the analysis is based on point patterns or spatial autocorrelation.

In all the analyses that are described below, care must be taken to distinguish the pattern of the plants or plant parts (e.g., leaves) from the disease-

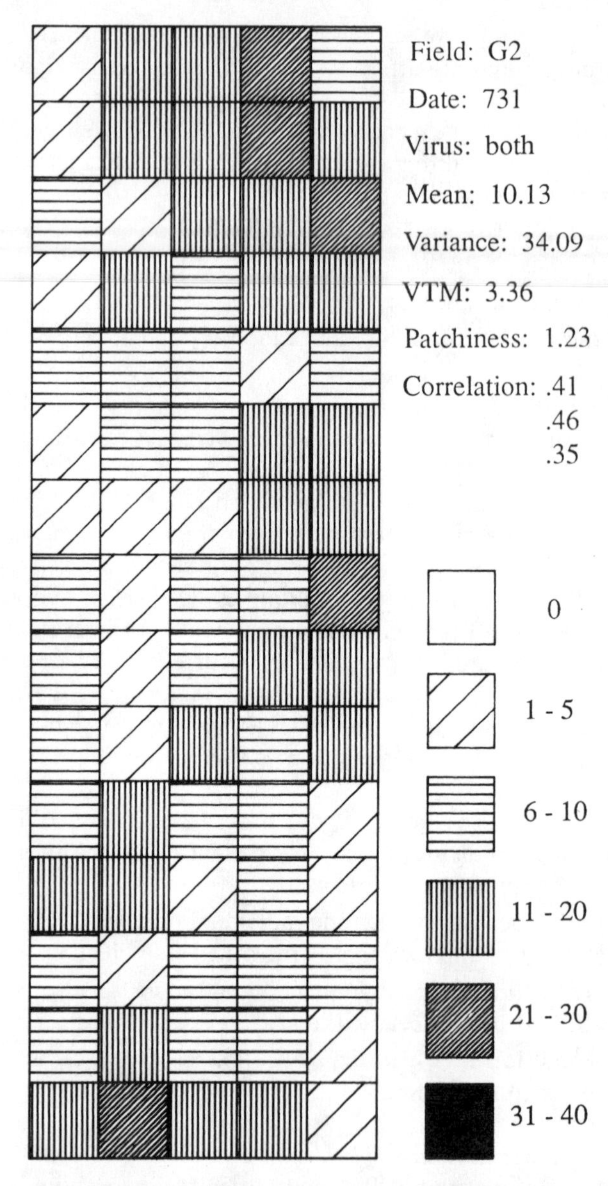

Figure 2 Spatial pattern of tobacco plants infected by tobacco vein mottling virus and tobacco etch virus. Each quadrat consists of 40 plants. Tobacco rows run down the page; each quadrat includes four rows. Numbers were grouped only for presentation, not for analysis. The variance-to-mean ratio (VTM) mean patchiness (m^*/m), and the first-order within-, across-, and diagonal-direction autocorrelations were significant at $P = 0.05$.

intensity pattern. With continuous (e.g., cover) crops or with plants in predetermined positions such as rows, this distinction should not cause difficulties. However, when plants in a given area have patterns ranging from random to highly aggregated (e.g., forest trees), one could misinterpret results from analysis of disease pattern. With the latter situation, it is advisable to assess the pattern of the crop and disease intensity.

Spatial Point Processes

If a field or other surface is divided into units such as quadrats and the number of points (e.g., diseased plants) is counted in each unit (Fig. 2), the spatial distribution of the points can be displayed and analyzed as a frequency distribution. When points on a surface have a random spatial pattern (see random distribution definition 1) and other restrictive assumptions are met, the frequency distribution of points is described by the Poisson distribution. The necessary assumptions include:

1. Maximum number of points a unit(e.g., quadrat) can contain equals a fixed value of n;
2. Expected number of points per unit is np, in which p is the probability of a location being occupied by a point;
3. There are no interactions between units nor a tendency for neighboring units to display similar "traits";
   and
4. np is small relative to n.

When n gets very large (approaches infinity), the probability of observing x points in a unit is given by

$$P_x = \frac{m^x e^{-m}}{x!} \tag{1}$$

in which $m = np$ and is a parameter representing the mean per quadrat.

One must be careful not to confuse a random spatial distribution with the Poisson statistical distribution. A spatial distribution or pattern is the way that points or values are arranged on a surface or in a field. A statistical distribution is the way in which "values are apportioned, with different frequencies, in a number of possible classes" (Pielou, 1977), that is, a histogram. Under certain conditions, a random spatial distribution will result in a Poisson statistical distribution. The Poisson and other statistical distributions are used to ascertain information on actual spatial patterns. Because of the strict assumptions listed above, one would not expect the Poisson distribution to occur very often. A chi-square (χ^2) goodness-of-fit test can be used to determine if there is a significant departure in the observed statistical distribution (say, at P = 0.05) from the theoretical Poisson.

Sometimes one is interested in assessing the distribution of "infections" and not infected plants, even though individual infections are not (easily) observable. This is especially true with soilborne pathogens. Gilligan (1983b) developed a very useful procedure for this situation, based on the Poisson distribution, in which the expected number of infected (or uninfected) roots is calculated for each of several inoculum densities. An interactive procedure to minimize the χ^2 goodness-of-fit statistic between observed and expected number of infected roots is used to estimate a parameter representing the "mean number of infections per root at the lowest, non-zero density of inoculum" (Gilligan, 1983b). There are two statistical limitations with Gilligan's procedure: (1) having expected numbers of infected roots <1 or even <5 for some inoculum levels renders the χ^2 test unreliable; and (2) information supplied by the replicates is not used in the procedure, only the total number of infected and uninfected plants per inoculum level. Despite these limitations, Gilligan's (1983b) procedure is an excellent method for evaluating randomness of infections.

With spatial point processes, there are two major extensions of the Poisson process that account for clustering of points. The first extension is for *true contagion,* sometimes called the *generalized Poisson process.* Here, clusters of points are randomly positioned on a plane, and the number of individuals per cluster is described by some statistical distribution such as the Poisson or logarithmic. As an example, consider the situation where a spore shower results in a random pattern of lesions in a field. After a generation, new lesions may be clustered around the original lesions. Now the clusters are randomly distributed within the field but the lesions per quadrat or sample have some other contagious or generalized distribution, such as the negative binomial or Neyman A.

Because of its relatively simple form, the negative binomial is used much more frequently than the Neyman A or other possible distributions. One formulation of the negative binomial can be written as

$$P_x = \binom{k + x - 1}{x} p^k (1 - p)^x \tag{2}$$

in which k and p are parameters.

The second expansion of the Poisson process is for *apparent contagion,* often called the *compound Poisson process.* Here, the units (quadrats) are dissimilar. Some may be more favorable for infection because of physical or crop factors, and other quadrats may be less favorable. Specifically, np is not constant for all units. Based on reasonable assumptions about the distribution of np ($= m$), the statistical distribution of the values can be represented by the negative binomial, Neyman A, and other distributions. These compound (as well as generalized) distributions represent many possible situations. In fact, the negative binomial can be derived in numberous ways; Boswell and Patil (1970) reviewed 14 derivations. One cannot determine the mechanisms that generated one of these compound or generalized distributions without additional information and assumptions. One possible approach is discussed later.

An extremely useful property of the Poisson distribution is that the variance (v) equals the mean (m), or equivalently, the variance-to-mean ratio (VTM) equals 1. Similarly, the compound and generalized distributions have a variance larger than the mean. One can calculate VTM for any set of point pattern observations and determine if VTM is significantly different from 1 with a chi-square test (Hoel, 1943):

$$\chi^2 = (N - 1)\mathrm{VTM} \tag{3}$$

in which N is the number of units (quadrats). One can reject the Poisson distribution (or spatial randomness under strict conditions) if the observed χ^2 is greater than a critical value from a chi-square table with $N - 1$ degrees of freedom.

A chi-square goodness-of-fit test can also be conducted for each statistical distribution after the appropriate parameters are estimated (Cliff and Ord, 1981). One could, then, quantify aggregation with an appropriate parameter of a distribution that fits the data observed. A computer program by Gates and Etheridge (1970) makes this operation fairly easy. Many plant pathogens and diseases have been analyzed in this way (e.g., see Campbell and Noe, 1985). Confusing aspects of this approach include situations when none of the commonly used statistical distributions fit the data or when two or more distributions fit the data equally well. For example, when analyzing the patterns of maize plants infected by maize dwarf mosaic virus, a third of the data sets could not be accurately fit by the negative binomial or Neyman A distributions, even though VTM was significantly ($P = 0.05$) greater than 1 (Madden et al., 1987a; Madden, unpublished). Another third of the data sets were fit very well by *both* the negative binomial and Poisson. Obviously, parameter estimation methods, sample size, and probably type I and II errors account for these results.

Fortunately, it is not always necessary to fit statistical distributions to spatial data. Calculations based on v and m produce indices that often are more than sufficient to assess point patterns. VTM is one of the often used indices. Instead of simply carrying out a chi-square test, one can use VTM to describe the degree of clustering; greater VTM indicates greater aggregation. Many other indices of aggregation have been proposed based on various theoretical arguments. Some of these, together with a possible interpretation in brackets, are presented here:

1. Index of cluster size, VTM − 1 [number of other points intimately associated (in a cluster) with a randomly chosen point]
2. Aggregation parameter from the negative binomial distribution, k, some-

times called the index of cluster frequency (number of clusters per quadrat)

3. Morisita's index, I_δ (scaled probability that two randomly chosen points belong to the same quadrat)
4. Lloyd's index of mean crowding m^* (VTM $- 1 + m$) (number of other points that are contained in a quadrat that contains a randomly chosen point; also degree of crowding experienced by an individual)
5. Lloyd's index of patchiness, m^*/m, ratio of mean crowding to mean density
5. Taylor's b, determined as the slope when log (v) is regressed on log (m), in which v's and m's are based on independent samples
7. Slope and intercept from regressing m^* on m obtained from independent samples (if linear, slope represents the patchiness of the pattern of clusters and the intercept represents the mean crowding of individuals within a cluster)

Despite being developed for different reasons and with different theoretical considerations (Morisita, 1962; Pielou, 1977; Taylor, 1984; Upton and Fingleton, 1985), there are many similarities among these indices. Of particular importance, it can be shown that

$$\frac{m^*}{m} \cong \frac{1}{k} + 1 \cong I_\delta \qquad (4)$$

These are equivalents, not equalities, but the agreement is quite close for most studies (Taylor, 1984). Arguments over whether to use m^*/m, k (or $1/k + 1$), or I_δ are fruitless since they give basically the same results.

Not listed here are the many indices based on distances. Upton and Fingleton (1985) can be consulted for the appropriate details and formulas for indices not presented here.

Spatial Autocorrelation Analysis

"Spatial autocorrelation is a property that mapped data possess whenever it exhibits an organized pattern" (Upton and Fingleton, 1985). Analysis under this topic is concerned with values (nominal, discrete, or continuous) measured or observed at specific locations. With the typical epidemiological studies considered here, the locations correspond to quadrats or sampling locations within a field. Coordinates of these quadrats are given by row and column coordinates i and j (Fig. 3). First-order correlations are determined by relating the value in each (i, j) quadrat with the values in the adjacent [e.g., $(i - 1, j)$] quadrats. Second- and higher-order correlations are based on quadrats separated by one or more quadrats (Fig. 3).

Comparisons can be made in several directions. One can consider spatial

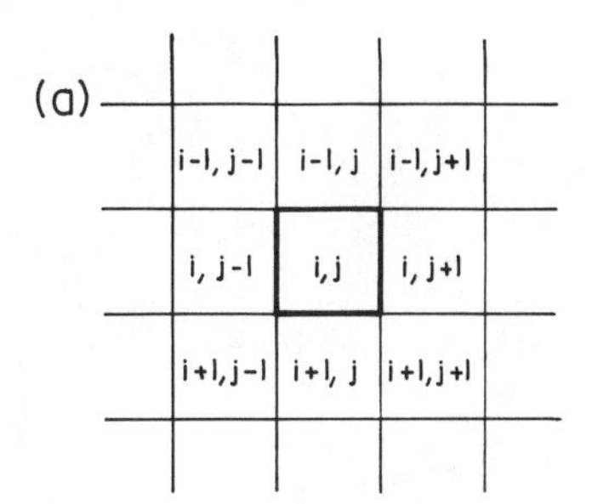

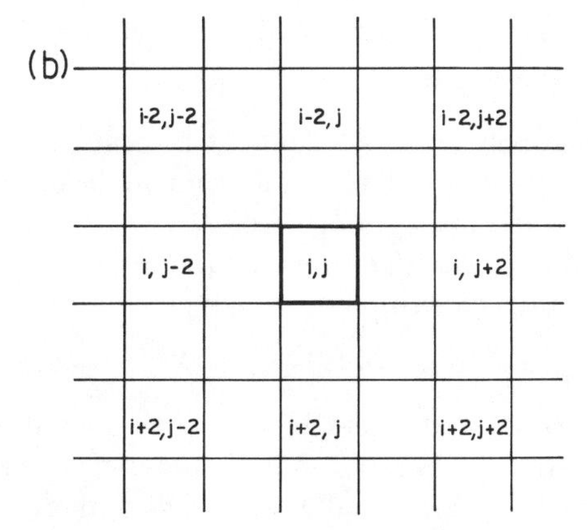

Figure 3 Portion of a field divided into quadrats. Parts (a) and (b) represent first- and second-order comparisons, respectively.

autocorrelation of values in which the quadrats share an edge ("rook's" case, using a chess analogy) (Fig. 4c). A restricted form of the rook's case is when comparisons are restricted either within or across rows (Fig. 4a and b). The "bishop's" case involves quadrats that share a point or corner; comparisons are in the diagonal direction (Fig. 4d). The "queen's" case includes the rook and bishop comparisons (combination of c and d in Fig. 4). The queen and rook cases are more common than the bishop's case.

When writing models, often it is sufficient to refer to any location (quadrat) with a single subscript. For instance, i and j can refer to any two (separate) locations in a field. The values at these locations (e.g., number of infected plants) are represented by X_i and X_j. Spatial autocorrelation models can be expressed in various forms (Bennett, 1979; Cliff and Ord, 1981). A common approach is to use an autoregressive model that can be written as

$$X_i = \beta \Sigma (W_{ij}X_j) + \mu + \xi_i \tag{5}$$

in which W_{ij} is a measure of spatial proximity of locations i and j, β is the spatial autocorrelation parameter, ξ_i is the error (disturbance) term with mean zero and constant variance, and μ is a constant. The errors can be thought to represent the combined effects of all unmeasured factors that influence X_i. As

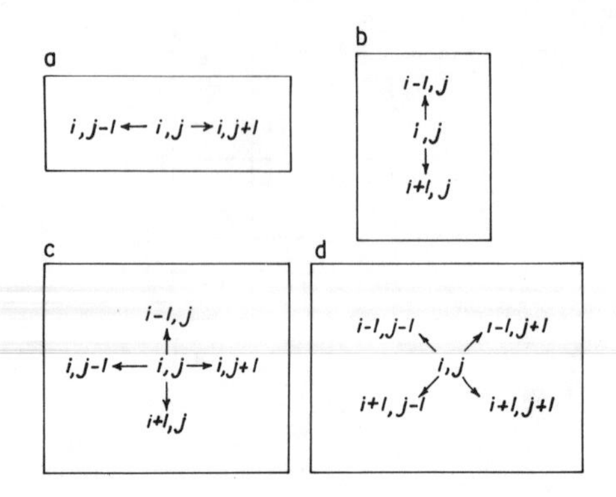

Figure 4 Some of the possible directional comparisons that can be made with spatial autocorrelation analysis: (a) across-row; (b) within-row; (c) rook's case; (d) bishop's case.

commonly used, W_{ij} equals 1 if locations i and j are in spatial proximity, and zero otherwise. The W's depend on the order (first, second, etc.), and direction (queen's, bishop's or rook's case) of the comparisons. The first-order rook case model (with a reversion to the two-subscript form to better show location) can thus be written as

$$X_{i,j} = \beta\,(X_{i-1,j} + X_{i+1,j} + X_{i,j-1} + X_{i,j+1}) + \mu + \xi_{i,j} \tag{6}$$

When $\beta > 0$, there is a positive spatial autocorrelation, indicating similar nearby X's. When $\beta < 0$, there is negative autocorrelation, indicating dissimilar nearby X's. Finally, when $\beta = 0$, there is no autocorrelation, and X_{ij} equals an independent, random variable $\mu + \xi_{i,j}$. The expected X equals μ. Often the overall mean routinely is subtracted from each X_{ij} prior to model development in order to obtain a zero-mean variable; then the constant μ is not necessary. There are several other formulations of the autoregressive model that can be found in Cliff and Ord (1981) and elsewhere. More compliated models, including those that contain functions of the proximal errors, can also be specified (Bennett, 1979).

Determining the most appropriate form of equation (5) and obtaining reasonable estimates of autoregressive model parameters is done with a cross-product statistic (ϕ). This statistic can be written in a general form as

$$\phi = \sum_{i}\sum_{j} (W_{ij}Y_{ij}) \tag{7}$$

in which W_{ij} is defined as before, Y_{ij} is a metric representing the closeness of two values at locations i and j, and i does not equal j. Two of the commonly used metrics for Y are $(X_i - X_j)^2$ and $(X_i - m)\,(X_j - m)$. When these two metrics are used for continuous data, equation (7) produces a scaled version of Geary's c and the spatial autocorrelation coefficient r, respectively (Upton and Fingleton, 1985). A special version of r is the Moran I statistic.

Bennett (1979) discusses the various methods of calculating spatial autocorrelation coefficients. The statistics c and I are analogous to the Durbin–

Watson statistic and the autocorrelation coefficient in time series analysis. Both statistics generally lead to the same conclusion, but I (or r) is the more commonly used. The order of the correlation is indicated by a subscript. For instance, a second-order autocorrelation would be indicated by r_2 (or I_2). For simplicity, first-order statistics often do not have the subscript 1. Significant departures from randomness can be determined with a standard normal statistic (Z) by subtracting the expected I for a random pattern ($E(I)$) from the calculated I, and then dividing by the standard deviation of $E(I)$. Autocorrelation coefficients can then be used as initial parameter estimates in a maximum likelihood procedure for fitting quite complicated spatial models to data. Even when one is not interested in fitting specific models to data, the autocorrelation coefficients are extremely valuable in quantifying the association of disease (or other measured variable) in proximal quadrats.

A special case of spatial autocorrelation analysis involves transects along a surface (i.e., a single row of contiguous quadrats). So-called *local variance* techniques can be used for assessing patterns in this situation (Upton and Fingleton, 1985). When the quadrats are replaced by single plants, one can use a form of equation (7) to calculate and analyze the number of *runs* of healthy and diseased plants. Because of certain features of transects and the binary data of single plants, special tests can be constructed for testing whether the observed number of runs differed from the expected number under randomization (Gibbons, 1976; Madden et al., 1982; Upton and Fingleton, 1985).

SPATIAL DISTRIBUTION AND DISEASE DYNAMICS

Although it is preferable to determine the spatial distribution of plant pathogens or diseased plants at several times, sometimes it is not possible, due to economical, logistical, or other reasons, to obtain data at more than a single time. Still, an investigator may wish to use these spatial data to understand the nature of disease progress over time. This is possible theoretically, but only if certain assumptions are met.

Generalized and Compound Distributions

When studying disease data as spatial point processes, one normally does not know if significant clustering is due to contagion (generalized distribution) or quadrats with unequal favorability for the disease (compound distribution). Contagion, as an example, would suggest that new lesions were originating from randomly distributed parent lesions. Such an intepretation regarding disease dynamics can be made theoretically if several rather strict assumptions are made (Cliff and Ord, 1981): (1) contagion (generalized distribution) implies small clusters (i.e., smaller than the quadrat size), so that the proportion of clusters overlapping a quadrat boundary is small; and (2) apparent contagion (compound distribution) implies that the variable mean per quadrat (m) varies slowly over space, so that the clusters often cover areas larger than the quadrat. Both assumptions are plausible if not always probable. One also must assume certain combinations of statistical distributions that make up the negative binomial (Cliff and Ord, 1981, pp. 90–96). To test for contagion, one calculates the parameters of the negative binomial for a range of quadrat sizes. Using the methods of Greig-Smith (1952) for determining the size of clusters, one starts with many small quadrats and successively combines these to form larger and larger quadrats. Variance calculations can indicate the approximate size of clusters in terms of number of original (small) quadrats. Calculating k and p of the negative binomial for each quadrat combination (i.e., lattice size), k will generally increase with lattice size and p will remain unchanged if contagion is the overriding spatial process (Fig. 5); k will be unchanged and p will decrease if a compound process is involved.

We tested this approach for tobacco fields in which plants were infected by two potyviruses, tobacco etch virus (TEV) and tobacco vein mottling virus (TVMV). Based on analysis of six fields at several times, in almost all cases k increased with lattice size, but p decreased (example in Fig. 5) (Madden et al., 1987b; Madden and Pirone, unpublished). Moreover, the maximum variance occurred when the fields were divided into 15 x 5 quadrats. One parameter suggested contagion but the other suggested a compound process! This confusing result may be due to the data arising from a mixture of two or more types of contagion, but more likely because, although the data did not have a true negative binomial distribution, that distribution was similar enough so that the chi-square goodness-of-fit test failed to detect a significant difference (Upton and Fingleton, 1985). Nevertheless, the results shown in Fig. 5 are quite consistent and we hope to eventually determine the cause.

Spatial Autocorrelation and Point Pattern Analyses

Spatial autocorrelation coefficients can provide some information on disease dynamics. Calculation of Moran's I (or r statistic) for low- and high-order comparisons can lead to some valuable conclusions (Table 1). Dispersal from a few sources or the existence of large clusters (patches, mosaics) are indicated by positive low-order I. Positive low-order I and a negative high-order I suggest a gradient of disease (or pathogen), possibly due to dispersal from a few foci. Small clusters would produce a negative low-order I when the clusters are smaller than the quadrats. Unique conclusions cannot be made based on spatial autocorrelations, but strong suggestions are possible. Sokal (1978) or Cliff and Ord (1981) should be consulted for a more detailed discussion. With tobacco and maize viruses, we typically found high and significant ($P = 0.05$) first-order spatial autocorrelaions, and lower, nonsignificant higher-order correlations (see the example in Fig. 6) (Madden, Louie and Knoke, unpublished; Madden et al., 1987b).

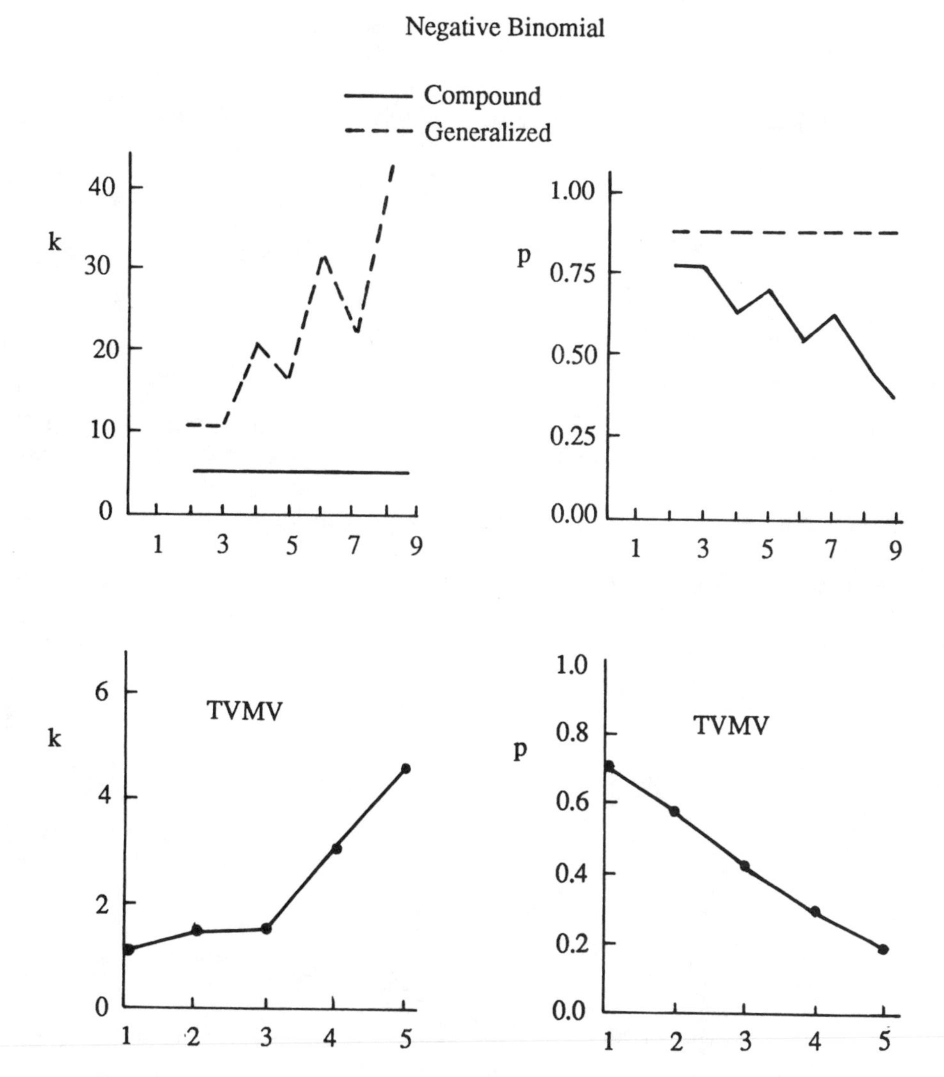

Figure 5 Upper part: a theoretical relationship between the parameters of the negative binomial distribution (k and p) and quadrat size (lattice). Lower part: actual values of k and p for tobacco vein mottling virus (TVMV)–infected plants.

Nicot et al. (1984) recently criticized the use of point pattern methods to interpret spatial distributions because the locations where disease occurs are not used in the analysis. Even though point pattern processes are based on the location of points, unless one is using distance methods, only the counts per sample (quadrat) are actually analyzed. Nevertheless, point patterns and autocorrelation analyses can be combined to provide valuable information on spa-

TABLE 1 Interpretation of Spatial Autocorrelation Coefficients

Sign of Autocorrelation	Order of Autocorrelation (Lag)	
	Low	High
Positive	(1) Dispersal from a few sources (2) Large favorable patches (3) Gradient	(1) Symmetrical surfaces (2) Patchy arrangement
Negative	(1) Heterogeneous area (2) Small patches	(1) Gradient

Source: Based on Cliff and Ord (1981).

tial processes (Table 2) (Cliff and Ord, 1981). Making the two numbered assumptions as above, one could conclude that contagion was the main spatial process if the negative binomial distribution fit the data or an index of aggregation indicated clustering, but Moran's I was *not* different from zero. Apparent contagion would have the same point pattern result, but I would be *greater* than zero. These results are all dependent on quadrat size (Cliff and Ord, 1981; Upton and Fingleton, 1985). When there is no obvious partitioning of a field (such as individual trees), one should use a form of Grieg-Smith's analysis to determine a reasonable number (size) of quadrats.

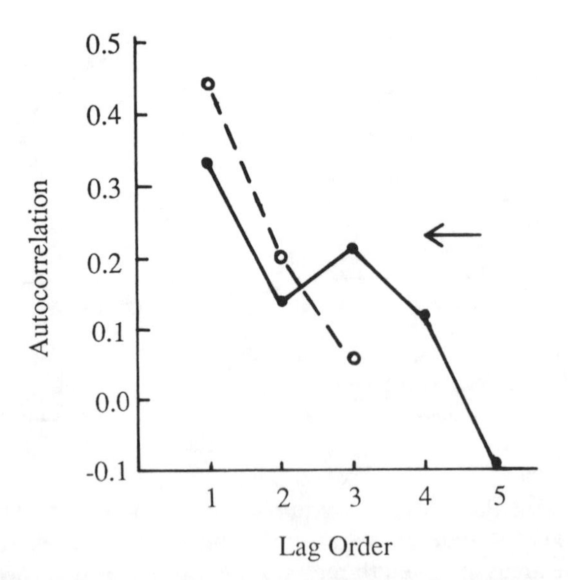

Figure 6 Spatial correlograms for autocorrelations within (•———•) and across (○— — —○) rows of tobacco plants infected by tobacco vein mottling virus. Results based on a field divided into 15 × 5 quadrats. Arrow indicates significance at $P = 0.05$.

TABLE 2 Interpretation of Combined Results for Spatial Autocorrelation and Point Pattern Analysis

	Statistical Distribution	
Spatial Autocorrelation	Simple Poisson	Generalized or Compound[a]
Not detected	Simple Poisson	True contagion
Detected[b]	Apparent contagion?	Apparent contagion

[a]For example, the negative binomial; more generally, when the variance is significantly greater than the mean.

[b]Correlation significantly different from zero.

Source: Based on Cliff and Ord (1981).

Combined results for the TVMV example would suggest apparent contagion (Figs. 5 and 6), with fairly large clusters of infected plants, and dispersal from a relatively few number of sources. Overall evidence for the field in Fig. 2, based on a single time, is that new infected plants are not *tightly* clustered around initial virus sources. Additional information on disease dynamics requires information for more than a single time (Madden et al., 1987b).

CHANGE IN PATTERNS OVER TIME

During an epidemic when pathogen and disease levels are increasing, one expects changes in some aspects of spatial distribution. Numbers of infected plants may increase more in some quadrats than in others. Similarly, size of clusters and numbers of individuals/cluster will be expected to depend on mean density. Number and length of runs of infected plants, also, will not be static.

Nonstancy of Aggregation Indices

Despite the expectations noted above, many basic and applied ecologists have attempted to describe the spatial distribution of many different organisms as being constant. This misdirected effort probably stems from early work with the negative binomial distribution in which a constant or common k parameter (k_c) was observed for some data sets with varying means. Theoretical work was then conducted to estimate k_c statistically from several data sets (Anscombe, 1948; Bliss and Owen, 1958; see the thorough review in Taylor, 1984). Also, it has been shown theoretically that k is not changed when deaths are *random* in a population. Taylor et al. (1979) questioned what the latter result is thought to prove, other than k will not change *if* mortality is spatially random and independent of density.

Since at least 1949, the negative binomial *k* has been known to vary with mean density (Anscombe, 1949). Even Bliss (1971), an early proponent of a common *k*, later indicated that there was strong evidence against a constant value. Still, many researchers, including plant pathologists, have attempted to determine a common *k*, either to characterize the distribution of an organism or to develop a sampling scheme (e.g., Strandberg, 1973; review by Taylor, 1984).

If *k* is not constant for changes in mean density, then by equation (4), neither are I_δ and m^*/m. Even though both indices can be developed theoretically as being independent of the mean or of random deaths, meeting all the assumptions necessary to reach this conclusion is unlikely. Taylor (1984) has additional arguments against the constancy of these aggregation indices.

There is no reason to expect spatial autocorrelation coefficients to remain constant during an epidemic or any other dynamic process. Cliff and Ord (1981), Haggett et al. (1977), Upton and Fingleton (1985), and others show numerous examples where spatial patterns change drastically over time, including detailed data on human epidemics. These results are based on nominal (presence–absence), count, and continuous data, using many different forms of autocorrelation statistics.

Fungal and Bacterial Examples

Plant pathologists seldom have attempted to characterize the change in spatial distribution as an epidemic progresses. Even if the data were collected and analyzed properly, detailed results have seldom been presented.

In studying the distribution of powdery mildew of wheat, Rouse et al. (1981) observed large changes in the negative binomial *k* over time. For example, *k* for the number of infected leaves went from 0.9 to 18.3 over 4 weeks in which the mean increased from 4.4 to 13.4 per quadrat. Poushinsky and Basu (1984) measured the aggregation of plants infected by *Pseudomonas syringae* pv. *glycinea* using distance methods. They found a general decrease in aggregation as disease incidence increased (see their Table 2). Griffin and Tomimatsu (1983) detected a negative relationship between m^*/m and the number of *Cylindrocladium crotalariae* infections per meter of peanut root. Campbell (1986) assessed the levels of *Macrophomina phaseolina* in sandy loam soil over 3 years. He found that aggregation measured with I_δ increased after planting during each year. Using runs analysis, Ferrin and Mitchell (1986b) found a consistent increase in the percentage of tobacco plots with a clustered pattern of plants killed by *Phytophthora parasitica* var. *nicotianae*. Reynolds et al. (1987) found that disease incidence of strawberry leather rot (caused by *Phytophthora cactorum*) became more aggregated over time. Results were based on a spatial autocorrelation analysis (rook's case) of the number of tagged cymes in each quadrat with one or more infected fruit. On the other hand, Thal and Campbell (1986) found little evidence for a systematic change in

VTM, I_δ, or Taylor's *b* over time for alfalfa leaf spot distributions. However, disease severity changed very little over time in the sampled fields.

Viral Disease Examples

Madden et al. (1987a) studied the dynamic nature of the aggregation of maize dwarf mosaic virus (MDMV)–infected maize plants at three locations in Ohio. Patchiness was determined by dividing fields into 36 continguous quadrats of 100 plants each and counting the number of infected plants in each quadrat. Runs of infected and healthy plants were calculated by sequentially moving up and down rows of the entire field and treating the last plant of row *i* as continuous with the first plant of row *j*. Patchiness and runs declined in 9 and 10 of 13 plots, respectively (e.g., Fig. 7). Patchiness results usually indicated clustering early in the epidemic and randomness (or regularity) later (Madden et al., 1987a). Runs analysis results usually were just the opposite. As numbers of infected plants increased, there were long runs of infected plants but a random or regular pattern of infected plants per quadrat. In addition to showing the differences between point patterns and spatial autocorrelation, these results demonstrate the effect of sample unit size on spatial distribution. Although less common, other changes in m^*/m and runs were observed (Fig. 8). Working with the same disease, Scott (1985) also found limited evidence for a positive relationship between disease incidence and degree of clustering based on runs analysis.

The spatial pattern of virus-infected tobacco plants was assessed with runs analysis by using a different approach. Each row was analyzed separately,

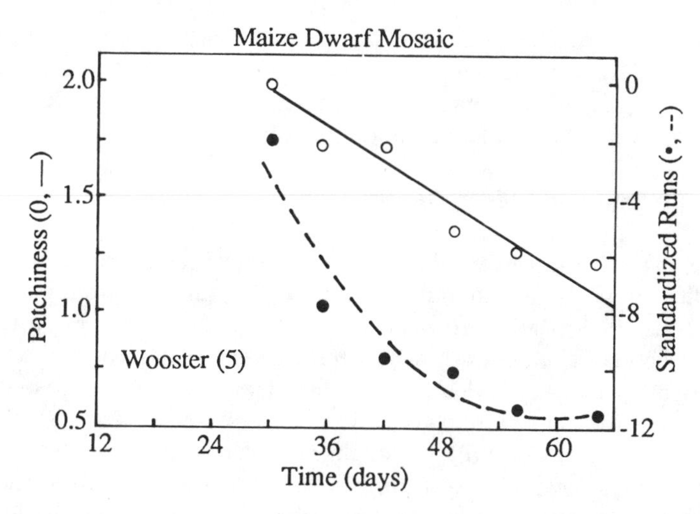

Figure 7 Mean patchiness and standardized runs versus time for maize plants infected with maize dwarf mosaic virus.

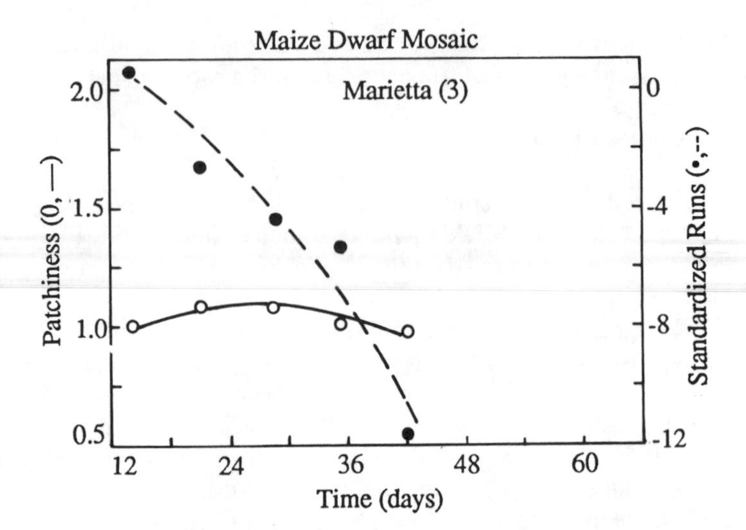

Figure 8 Mean patchiness and standardized runs versus time for maize plants infected with maize dwarf mosaic virus.

and the percentage of rows in each field that had a clustered pattern was determined (Madden et al., 1987b). In general, the percentage increased with time during the early part of the epidemics and then declined. If m was much below its maximum level, the percentage increased throughout the epidemic (Fig. 9). Patchiness also changed over time with the tobacco virus disease epidemics, increasing to some maximum fairly early in the epidemics, and then declining (Madden et al., 1987b) (e.g., Fig. 10). First-order, rook's case spatial autocorrelations were also calculated using the computer program by Modjeska and Rawlings (1983). As indicated by this statistic, aggregation generally increased throughout much of the tobacco virus epidemics (Fig. 10) and declined only when m was very close to its maximum possible level.

Aggregation and Mean Density

The many forms of variation in aggregation with time might suggest that there is little hope in understanding and predicting this variation. Fortunately, Taylor et al. (1979) shed a great deal of light on the subject when they presented a theoretical relationship between aggregation (measured as m^*/m, $1/k + 1$, or I_δ) and mean density (Fig. 11). The relationship is based on the postulate that the change in log (v) with change in log (m) for a given species is a constant (i.e., Taylor's b). Although this might seem as unlikely as assuming a constant index such as m^*/m, Taylor and coworkers have shown the widespread appropriateness of the "power law" for v and m (Taylor, 1961, 1971; Taylor et al., 1978, 1979). Numerous data sets representing more than

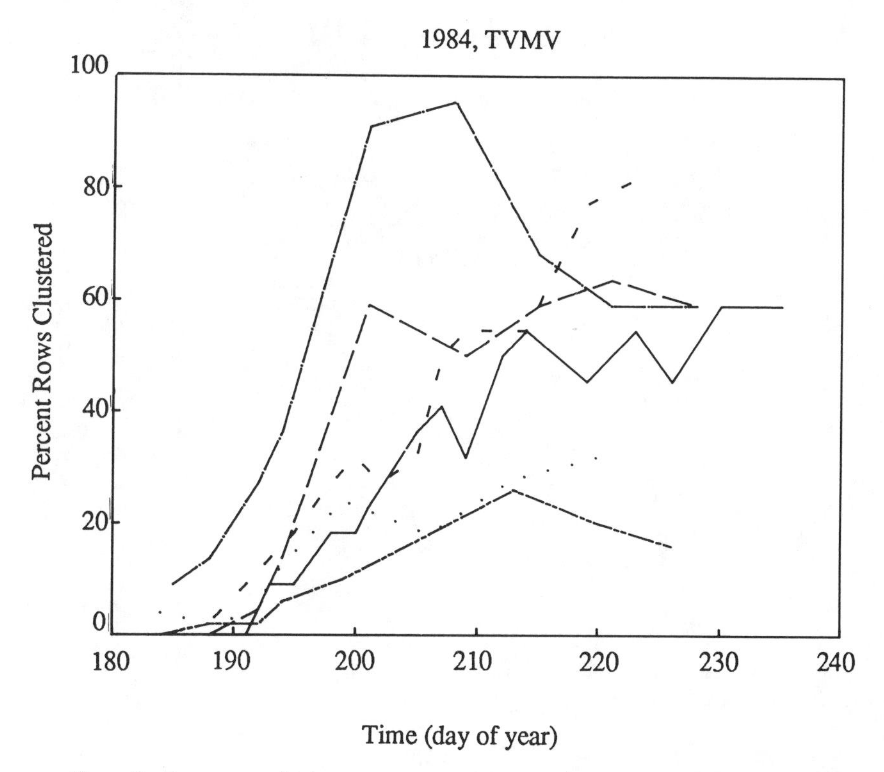

Figure 9 Percentage of tobacco rows (out of 22) in which a runs analysis indicated significant clustering ($P = 0.05$) of tobacco vein mottling virus-infected plants in six fields.

100 species, mainly insects, were analyzed and found to be precisely described by regressing log (v) on log (m). The appropriateness of Taylor's power relation for describing the distribution of insects, including virus vectors (Walker et al., 1984), is truly remarkable. Figure 11 is based on a slope b being between (and not including) 1 and 2; this is the typical range in the literature. If $b = 1$, the line would be horizontal; if $b > 2$, the line is monotonically increasing. It is important to note that only a portion of the curve might be realized with a given data set (Taylor et al., 1979). For example, all the samples might be taken when mean density [log (m) to be more precise] is in the midrange, resulting in aggregation indices that show little or no change. Other times, only relatively low or high densities are sampled and aggregation will either increase or decrease, respectively.

Mean disease levels increase as epidemics progress. Therefore, the functional relationsip of Taylor et al. (1979) could have great application for studying the dynamic aspects of disease and pathogen distributions. At this time, however, the appropriateness of Taylor's power law to plant pathogens is not

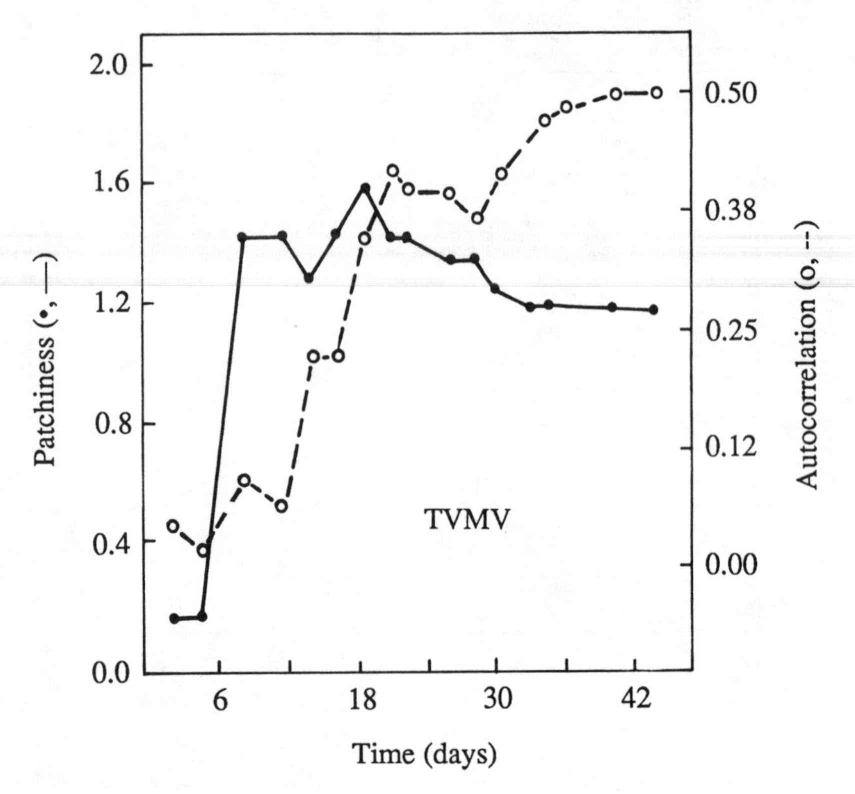

Figure 10 Mean patchiness and first-order spatial autocorrelation versus time for tobacco plants infected by tobacco vein mottling virus.

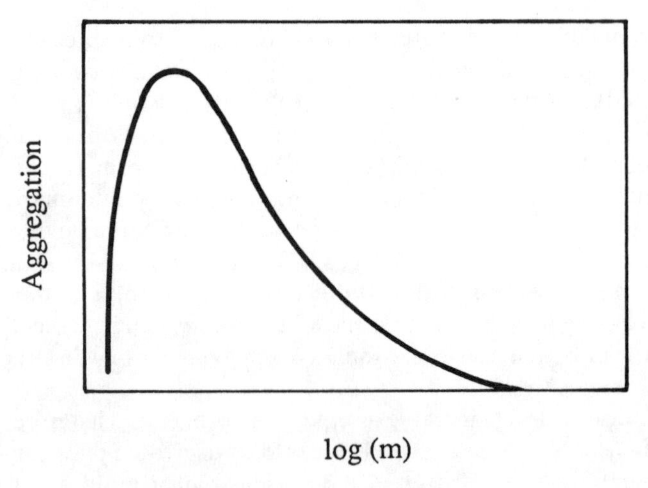

Figure 11 Aggregation (m^*/m, $1/k + 1$, or I_δ) versus log of mean density for the situation when Taylor's b is greater than 1 but less than 2.

clear. Thal and Campbell (1986) found a good fit when regressing log (v) and log (m) for alfalfa leaf spot data. Madden et al. (1987a) were less successful, finding a nonsignificant relationship between the variance and the mean in 3 of 13 plots with MDMV-infected maize. In other plots, there was a disagreement between Taylor's results and the aggregation results for each time. With tobacco viruses, we found a good relationship in some fields, especially if mean density was fairly low (Madden et al., 1987b) (Fig. 12). In many fields, however, the variance did not increase at high disease levels (Fig. 12).

A complicating factor when studying plant disease distributions is the inherent aspect of a maximum disease level. The variance may not increase as the mean approaches the maximum, in fact, v may even decrease. This type of data was not considered by Taylor et al. (1978,1979) in postulating their power law. Despite this complication, the power law should be explored further for describing the aggregation of plant diseases, especially when densities are not near this maximum. Methods could also be developed to correct for maximum disease level prior to the spatial distribution analysis.

EPIDEMIC CHARACTERISTICS AND SPATIAL DISTRIBUTIONS

During the course of a plant disease epidemic, the disease level and its spatial distribution will vary. Generally, disease level will increase and, as in the preceding section, distribution will change in a more complicated, less predictable fashion. It would be useful to know if disease intensity or its rate of change

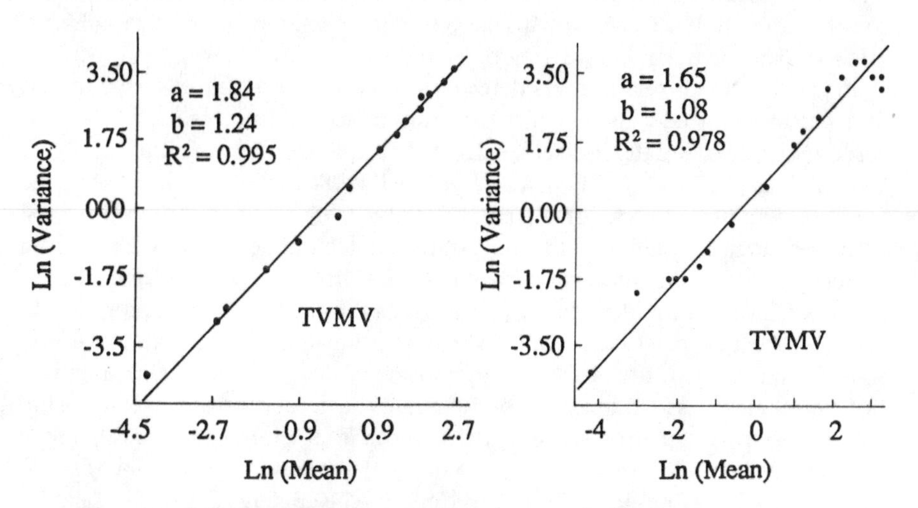

Figure 12 Relationship between logarithm of the variance and logarithm of mean density for tobacco plants infected by tobacco vein mottling virus in two fields. *a* is the intercept and *b* the slope.

at some time t had any influence on the disease pattern at time $t + 1, 2, 3, \ldots, n$. Additionally, it would be informative to know if highly aggregated or randomly distributed infected plants at time t would result in a greater increase in disease intensity at a later time. Put more generally, one can ask: Do the temporal aspects of an epidemic have any influence on the spatial aspects? Conversely, one can ask: Do the spatial aspects of an epidemic influence the temporal aspects? These questions have not been well answered by epidemiologists.

In studying epidemics of watermelon mosaic virus 2 (WMV-2) on muskmelon, Gray et al. (1986) assessed the distribution of infected plants and found a high degree of aggregation throughout disease epidemics of an early season planting. The epidemics were all well described by the logistic model. For a late season planting, distributions generally were random until late in the epidemics; these epidemics were well described by a simple linear model. Gray et al. (1986) concluded that disease increase in early and late season plantings was due mainly to within-field and external virus sources, respectively. Early season plantings were associated with relatively low final disease incidence (mostly <0.5), whereas late season plantings had final incidence levels greater than 0.9.

Campbell et al. (1984) found a random sequence of tobacco plants infected with *Phytophthora parasitica* var. *nicotianae* using runs analysis, suggesting no spread from plant to plant (Madden et al., 1982). They found no relation between aggregation and the form of the disease progress curve nor, apparently, other temporal characteristics of the black shank epidemics. Working with the same pathosystem, Ferrin and Mitchell (1986a) found a negative correlation between aggregation of initial inoculum and disease incidence for the latter half of the growing season with a highly susceptible cultivar. Highest percent mortality was associated with the lower levels of inoculum aggregation. No correlation was detected for a less susceptible cultivar. Inoculum was represented as propagules per gram of soil and aggregation was measured with Lloyd's patchiness. Apparent discrepancies between the results of Campbell et al. (1984) and Ferrin and Mitchell (1986a) probably are due to the measurement and analysis of different variables (e.g., aggregation of infected plants or inoculum) and the different statistical techniques used for evaluating patterns. It would be interesting to conduct a follow-up study in which aggregation of inoculum and diseased plants throughout the season is monitored.

Madden et al. (1987a) explored the relationship between temporal and spatial characteristics of maize dwarf mosaic epidemics. Measuring aggregation with patchiness (36 quadrats of 100 plants each) and runs, no relationship was found between initial aggregation and initial disease level (y_0), mean weighted rate of disease increase (ρ), shape of the disease progress curve, or final disease level (y_f). Patchiness at the end of the epidemics was significantly negatively correlated ($P < 0.05$) with y_o and ρ (e.g., Fig. 13). Higher initial disease and rate of increase resulted in less aggregation of infected plants.

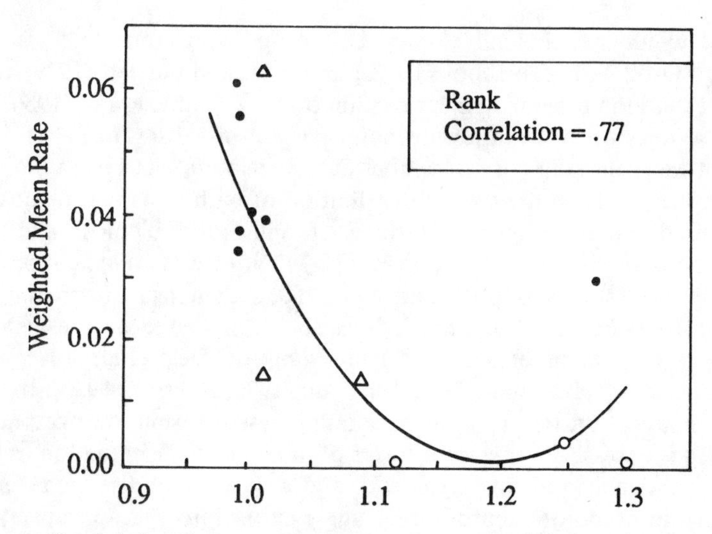

Figure 13 Relationship between weighted mean rate of disease increase and final mean patchiness for maize dwarf mosaic virus-infected maize plants in 13 fields. Line represents a quadratic equation fit to the data. Symbols: closed circles, logistic disease increase; open circles, monomolecular; triangles, Gompertz.

The epidemics best described by the monomolecular model had a higher final patchiness than most of the other epidemics. Unfortunately, there were not enough epidemics adequately described by each disease progression model to reach a conclusion regarding the type of disease increase and aggregation. There was no correlation between the shape of the disease progress curves, as represented by the estimated Weibull shape parameter, and final patchiness.

Considerably more research relating temporal and spatial epidemic characteristics is needed. As yet, it is not possible to formulate broad principles regarding the relationship, if any, between these major components of plant disease epidemics.

DISCUSSION AND CONCLUSIONS

Prior to 1980, few plant pathologists were interested in quantifying spatial distributions of pathogens and diseases. Since then, a considerable amount of research has been conducted on describing and understanding spatial patterns. The recent article by Campbell and Noe (1985) summarizes the many studies on soilborne pathogens and the diseases they cause. Plant pathologists wishing to start this type of research now are fortunate because of the publication of some very good books and articles that clearly describe the statistical methods and the proper interpretation of results (Bennett, 1979; Cliff and Ord, 1981; Elliott, 1977; Ripley, 1981; Ruesink, 1980; Sokal, 1978; Upton and Fingleton,

1985). I would recommend reading Upton and Fingleton (1985) first, then Cliff and Ord (1981). The books by Ripley (1981) and Bennett (1979) are more difficult than the others. Articles by Elliott (1977) and Ruesink (1980) are excellent sources of information on point pattern processes. Sokal (1978) provides an excellent summary of spatial autocorrelation analysis in biology.

Information on the spatial distribution of pathogens and plant diseases can be used in many ways. The influence of cultural, biological, and physical factors on plant diseases can be assessed by their effects on spatial patterns at one or more times (Campbell and Noe, 1985). Sample size and the pattern through the field that the sampler takes should be based on the distribution of the pathogen or other organism being sampled (Delp et al., 1986; Goodell and Ferris, 1981; Hau et al., 1982; Lin et al., 1979; Ruesink, 1980). In general, greater aggregation requires larger sample sizes. Information on spatial patterns also is necessary for certain forms of modeling and simulation. Although much of epidemic modeling assumes a random pattern of disease, recent work suggests the need for incorporating aggregation into the appropriate equations (Gilligan, 1983a; Mundt et al., 1986; Waggoner and Rich, 1981).

It was shown in this chapter that measurements of aggregation at a single time provide, among other things, valuable information on the number and size of clusters, crowding experienced by an individual, as well as the association of disease values at various locations within a field. Under strict assumptions, one can reach additional conclusions concerning disease dynamics. Without these assumptions, such conclusions certainly would be suspect.

Aggregation of plant diseases and plant pathogens, like most other organisms, vary with mean density and therefore usually with time. This dynamic aspect of disease distributions can be observed no matter how aggregation is assessed, as point patterns or with spatial autocorrelation analysis. Epidemiological data suggest that there are some important correlations between spatial and temporal characteristics of plant disease epidemics. Although detailed studies along these lines are limited, it is hope that these relationships are considered in future investigations.

A more fundamental approach to consider the spatial-temporal components of an epidemic is needed. Instead of correlating spatial and temporal measurements or describing the change in aggregation with empirical regression equations, fundamental spatial-temporal models, ideally, should be strived for.

Logistic Increase and Disease Aggregation

Waggoner and Rich (1981) already have made an ingenious contribution to this area by expanding the logistic equation to account for clustering of lesions or infections. They used the negative binomial k parameter to alter the so-called correction factor, $1 - x$, of the logistic model. Increased clustering would lead to a lower absolute rate of disease increase. Unfortunately, there are two practical problems with this approach. First, their expanded equation(s) cannot be integrated analytically; only numerical solutions are possible. Second, despite their claims to the contrary, there is ample evidence that aggregation (and thus k) changes with disease level. The first problem can easily be overcome with computer programs to integrate differential equations. However, the second problem presents a more difficult obstacle for modeling the increase of clustered diseases. Perhaps another expansion of the logistic equation using Taylor's b (which incorporates the effect of mean density on the variance) could be developed. Unfortunately, it is not readily seen how this would be done. Despite the difficulties mentioned here on the utility of the revised logistic equation, Waggoner and Rich (1981) have made an outstanding theoretical contribution to this area.

Spatiotemporal Autocorrelation Analysis

A different approach for fundamentally describing spatiotemporal epidemic characteristics involves an expansion of the spatial autocorrelation model [equation(5)]. Previously it was shown how disease at location i can be related to disease at one or more proximal j locations. Disease at time t and location i (X_i,t) can be related to disease at the same location but at time $t - 1$ using the simple temporal autoregressive model

$$X_{i,t} = \beta X_{i,t-1} + \xi_{i,t} \qquad (8)$$

One can now consider disease in a field from a purely spatial (Fig. 14, line B) or purely temporal (Fig. 14, line A) setting. A generalized expansion of equations (5) and (8) can be written to account for spatiotemporal processes:

$$X_{i,t} = f(X_{i,t-k}, X_{j,t-k}) + \mu + \xi_{i,t} \qquad (9)$$

in which i does not equal j and $k > 0$. One specific formulation can be written for $t = 1$ (lag one) as

$$X_{i,t} = \beta_1 X_{i,t-1} + \beta_2 \Sigma (W_{ij} X_{j,t-1}) + \mu + \xi_{i,t} \qquad (10)$$

in which β_1 and β_2 are two unknown parameters. Disease at time t in quadrat (location) i is dependent on disease at the previous time in the same location and also on disease in the neighboring quadrats at $t - 1$ (Fig. 14, line C). Such a model could be used to determine the relative importance of disease within and outside a given quadrat on disease increase in the quadrat.

With epidemics one does not expect X to be stationary (i.e., time invariant), a requirement for this type of analysis. One needs to calculate first differences to remove the trend in X over time. Define $Z_{i,t} = X_{i,t} - X_{i,t-1}$ as the first difference and substitute $Z_{i,t}$ for $X_{i,t}$ in equations (8)–(10). All analyses are then performed on the $Z_{i,t}$ instead of $X_{i,t}$.

Generalized cross-product statistics such as r (r_{st} = autocorrelation at spatial lag s and temporal lag t) and Moran's I (I_{st}) can be used to analyze

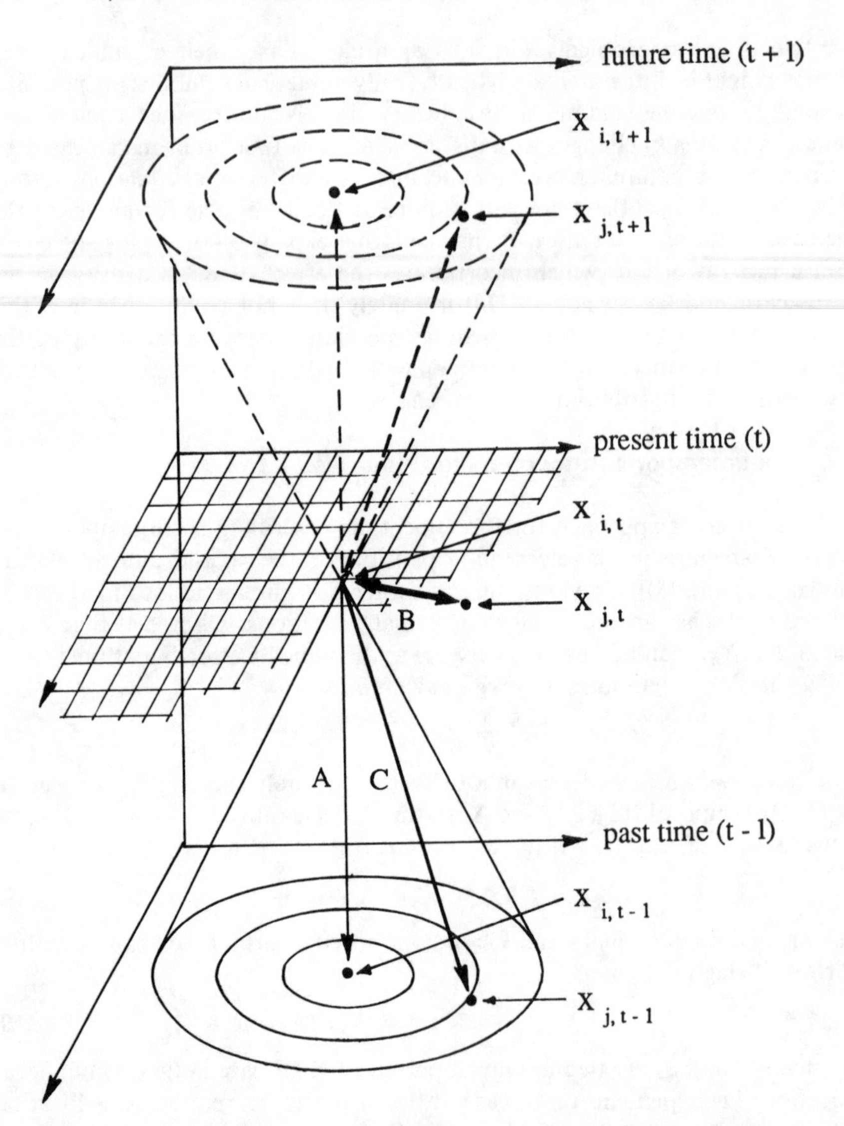

Figure 14 Autocorrelations in time and space. **A,** pure temporal autoregression; **B,** pure spatial autoregression; **C,** mixed spatial-temporal autoregression.

spatiotemporal data. As expected, complexity of this approach is much greater than either the purely temporal or spatial approaches. Model evaluation is difficult. Moreover, computer programs for performing spatiotemporal or even purely spatial analysis are not as readily available as other statistical programs. Despite these limitations, analysis of this type has great promise for increasing our understanding of plant disease dynamics. We have developed the necessary computer software for calculating spatiotemporal autocorrelation coefficients for a range of spatial and temporal lags (Reynolds and Madden, 1987). Preliminary results of viral and fungal disease data suggests that plant disease epidemics can be represented by relatively simple spatiotemporal statistical models. The overall goal of this research, which can be stated with a quotation from Cliff and Ord (1981), is "to unravel the complex patterns of autocorrelation in both time and space to gain some insight into the functional dependencies between areas implied by the presence of autocorrelation."

ACKNOWLEDGMENTS

The cooperation of J. K. Knoke, R. Louie, T. P. Pirone, and B. Raccah in conducting the virus disease research is greatly appreciated. I thank C. L. Campbell, L. J. Francl, and K. M. Reynolds for their comments on an early draft of this chapter.

Salaries and research support provided by state and federal funds appropriated to the Ohio Agricultural Research and Development Center, The Ohio State University. Journal Article No. 239-287.

REFERENCES

ANSCOMBE, F. J. 1948. On estimating the population of aphids in a potato field. Ann. Appl. Biol. 35:567–571.

ANSCOMBE, F. J. 1949. The statistical analysis of insect counts based on the negative binomial distributions. Biometrics 5:165–173.

BALD, J. G. 1937. Investigations on "spotted wilt" of tomatoes. III. Infection in field plots. Commonw. Aust. Counc. Sci. Ind. Res. Bull. 106.

BENNETT, R. J. 1979. *Spatial Time Series.* Pion Ltd., London.

BLISS, C. I. 1971. The aggregation of species within spatial units. Stat. Ecol. 1:311–335.

BLISS, C. I., and OWEN, A. R. G. 1958. Negative binomial distributions with a common *k*. Biometrika 45:37–58.

BOSWELL, M. T., and PATIL, G. P. 1970. Chance mechanisms generating the negative binomial distributions. *In* Random Counts in Scientific Work, Vol. 1 (ed. G. P. Patil), pp. 3–22. The Pennsylvania State University Press, University Park, Pa.

CAMPBELL, C. L. 1986. Spatial pattern dynamics of propagules of *Macrophomina phaseolina.* Phytopathology (Abstr.) 76:1129.

CAMPBELL, C. L., and NOE, J. P. 1985. The spatial analysis of soilborne pathogens and root diseases. Annu. Rev. Phytopathol. 23:129–148.

CAMPBELL, C. L. JACOBI, W. R., POWELL, N. T., and MAIN, C. E. 1984. Analysis of disease progression and the randomness of occurrence of infected plants during tobacco black shank epidemics. Phytopathology 74:230–235.

CLIFF, A. D., and ORD, J. K. 1981. *Spatial Processes: Models and Applications.* Pion Ltd., London.

DELP, B. R., STOWELL, L. J., and MAROIS, J. J. 1986. Field runner: a disease incidence, severity, and spatial pattern assessment system. Plant Dis. 70:954–957.

ELLIOTT, J. M. 1977. Some Methods for the Statistical Analysis of Samples of Benthic Invertebrates, 2nd ed. Freshwater Biol. Assoc. Sci. Publ. 25. Ambleside, Cumbria, England.

FERRIN, D. M., and MITCHELL, D. J. 1986a. Influence of initial density and distribution of inoculum on the epidemiology of tobacco black shank. Phytopathology 76:1153–1158.

FERRIN, D. M., and MITCHELL, D. J. 1986b. Influence of soil water status on the epidemiology of tobacco black shank. Phytopathology 76:1213–1217.

GATES, C. E., and ETHERIDGE, F. G. 1970. A generalized set of discrete frequency distributions with Fortran program. Math. Geol. 4:1–24.

GIBBONS, J. D. 1976. Nonparametric Methods for Quantitative Analysis. Holt, Rinehart and Winston, New York.

GILLIGAN, C. A. 1983a. Modeling of soil-borne pathogens. Annu. Rev. Phytopathol. 21:45–64.

GILLIGAN, C. A. 1983b. A test for randomness of infection by soilborne pathogens. Phytopathology 73:300–303.

GOODELL, P. B. and FERRIS, H. 1981. Sample optimization for five plant parasitic nematodes in an alfalfa field. J. Nematol. 13:304–313.

GRAY, S. M., MOYER, J. W., KENNEDY, G. G., and CAMPBELL, C. L. 1986. Spatial and temporal changes in spread of watermelon mosaic virus 2 caused by suppressive virus and aphid resistance. Phytopathology 76:1254–1259.

GREGORY, P. H. 1973. The Microbiology of the Atmosphere, 2nd ed. Halsted Press, New York.

GREIG-SMITH, P. 1952. The use of random and contiguous quadrats in the study of the structure of plant communities. Ann. Bot. 16:293–316.

GRIFFIN, G. J., and TOMIMATSU, G. S. 1983. Root infection pattern, infection efficiency, and infection density-disease incidence relationships of *Cylindrocladium crotalariae* on peanut in field soil. Can. J. Plant Pathol. 5:81–88.

HAGGETT, P., CLIFF, A. D., and FREY, A. 1977. Locational Analysis in Human Geography. Edward Arnold (Publishers) Ltd., London.

HAU, F. C., CAMPBELL, C. L., and BEUTE, M. K. 1982. Inoculum distribution and sampling methods for *Cylindrocladium crotalariae* in a peanut field. Plant Dis. 66:568–571.

HELTSHE, J. F., and RITCHEY, T. A. 1984. Spatial pattern detection using quadrat samples. Biometrics 40:877–885.

HOEL, P. G. 1943. On indices of dispersion. Ann. Math. Stat. 14:155–162.

LIN, C. S., POUSHINSKY, G., and MAUER, M. 1979. An examination of five sampling methods under random and clustered disease distributions using simulation. Can. J. Plant Sci. 59:121–130.

MADDEN, L. V., LOUIE, R., ABT, J. J., and KNOKE, J. K. 1982. Evaluation of tests for randomness of infected plants. Phytopathology 72:195–198.

MADDEN, L. V., LOUIE, R., and KNOKE, J. K. 1987a. Temporal and spatial analysis of maize dwarf mosaic epidemics. Phytopathology 77:148–156.

MADDEN, L. V., PIRONE, T. P., and RACCAH, B. 1987b. Analysis of spatial patterns of virus-diseased tobacco plants. Phytopathology 77:1413–1417.

MARCUS, R., FISHMAN, S., TALPAZ, H., SALOMON, R., and BAR-JOSEPH, M. 1984. On the spatial distribution of citrus tristeza virus disease. Phytoparasitica 12:45–52.

MINOGUE, K. P. 1986. Disease gradients and the spread of disease. *In* Plant Disease Epidemiology: Populations Dynamics and Management, Vol. 1 (ed. K. J. Leonard and W. E. Fry), pp. 285–310. Macmillan Publishing Company, New York.

MODJESKA, J. S., and RAWLINGS, J. O. 1983. Spatial correlation analysis of uniformity data. Biometrics 39:373–384.

MORISITA, M. 1962. I_δ-index, a means of dispersion of individuals. Res. Popul. Ecol. (Kyoto) 4:1–7.

MUNDT, C. C., LEONARD, K. J., THAL, W. M., and FULTON, J. H. 1986. Computerized simulation of crown rust epidemics in mixture of immune and susceptible oat plants with different genotype unit areas and spatial distributions of initial disease. Phytopathology 76:590–598.

NICOT, P. C., ROUSE, D. I., and YANDELL, B. S. 1984. Comparison of statistical methods for studying spatial patterns of soilborne plant pathogens in the field. Phytopathology 74:1399–1402.

PIELOU, E. C. 1977. Mathematical Ecology. John Wiley & Sons, Inc., New York.

POUSHINSKY, G., and BASU, P. K. 1984. A study of distribution and sampling of soybean plants naturally infected with *Pseudomonas syringae* pv. *glycinea.* Phytopathology 74:319–326.

REYNOLDS, K. M., and MADDEN, L. V. 1988. Analysis of epidemics using spatio-temporal autocorrelation. Phytopathology 78:240–246.

REYNOLDS, K. M., MADDEN, L. V., and ELLIS, M. A. 1988. Spatio-temporal analysis of epidemic development of leather rot of strawberry. Phytopathology 78:246–252.

RIPLEY, B. D. 1981 Spatial Statistics. John Wiley & Sons, Inc., New York.

ROUSE, D. I., MACKENIZE, D. R., NELSON, R. R., and ELLIOTT, V. J. 1981. Distribution of wheat powdery mildew incidence in field plots and relationship to disease severity. Phytopathology 71:1015–1020.

RUESINK, W. G. 1980. Introduction to sampling theory. *In* Sampling Methods in Soybean Entomology (ed. M. Kogan and D. C. Herzog), pp. 61–78. Springer-Verlag, New York, Inc., New York.

SCOTT, G. E. 1985. Nonrandom spatial distribution of aphid-vectored maize dwarf mosaic. Plant Dis. 69:893–895.

SOKAL, R. R. 1978. Spatial autocorrelation in biology. 1. Methodology. Biol. J. Linn. Soc. 10:199–228.

STRANDBERG, J. 1973. Spatial distribution of cabbage black rot and the estimation of diseased plant populations. Phytopathology 63:998–1003.

TAYLOR, L. R. 1961. Aggregation, variance and the mean. Nature (London) 189:732–735.

TAYLOR, L. R. 1971. Aggregation as a species characteristic. Stat. Ecol. 1:357–377.

TAYLOR, L. R. 1984. Assessing and interpreting the spatial distributions of insect populations. Annu. Rev. Entomol. 29:321–357.

TAYLOR, L. R., WOIWOD, I. P., and PERRY, J. N. 1978. The density dependence of spatial behaviour and the rarity of randomness. J. Anim. Ecol. 47:383–406.

TAYLOR, L. R., WOIWOD, I. P., and PERRY, J. N. 1979. The negative binomial as an ecological model and the density-dependence of *k*. J. Anim. Ecol. 48:289–304.

THAL, W. M., and CAMPBELL, C. L. 1986. Spatial pattern analysis of disease severity data for alfalfa leaf spot caused primarily by *Leptosphaerulina briosiana*. Phytopathology 76:190–194.

UPTON, G., and FINGLETON, B. 1985. Spatial Data Analysis by Example. John Wiley & Sons Ltd., Chichester, West Sussex, England.

VANDERPLANK, J. E. 1946. A method for estimating the number of random groups of adjacent diseased plants in a homogeneous field. Trans. R. Soc. S. Afr. 31:269–278.

WAGGONER, P. E., and RICH, S. 1981. Lesion distribution, multiple infection, and the logistic increase of plant disease. Proc. Natl. Acad. Sci. USA 78:3292–3295.

WALKER, G. P., MADDEN, L. V., and SIMONET, D. E. 1984. Spatial dispersion and sequential sampling of the potato aphid, *Macrosiphum euphorbiae* (Homopetra: Aphididae), on processing-tomatoes in Ohio. Can. Entomol. 116:1069–1075.

6 Diffusion and Spatial Probability Models for Disease Spread

K. P. Minogue

INTRODUCTION

The spatial properties of a spreading pathogen population are strongly influenced by the spatial properties of the mechanism by which the population is dispersed. An implication of this is that quantitative knowledge of dispersal should allow quantitative prediction of the spatial development of an epidemic. Such prediction is simplified if we have a clear idea of how dispersal can be parameterized in a useful and meaningful way. It is also helpful if we start with a sound theoretical understanding of the range of spatial behavior that a population can be expected to exhibit, and how the parameters of dispersal determine what kind of behavior is realized.

Two principal types of spread can be distinguished in pathogen populations. Individual foci tend to be more or less regular in form, with well-defined gradients, spreading outward at a constant velocity. Such spread has been aptly termed a "wave" (Fisher, 1937) and may be described by simple diffusion models. The formation of new foci, however, is a much less regular process, dependent on the random occurrence of individual events of low probability. Simple description is difficult, and the wave analogy in particular is usually inappropriate.

Note: Department of Botany and Plant Pathology, Purdue University, West Lafayette, IN 47907, US.

These two modes of spread do not, however, necessarily correspond to different mechanisms of dispersal, but merely represent arbitrary divisions of a continuum of behavior for which a single mechanism may be responsible. Any attempt to relate the developing spatial pattern of a population to dispersal should therefore account for both focal formation and focal expansion without requiring dual mechanisms. Current mathematical models of disease spread (see McCartney and Fitt, 1986; Minogue, 1986; Bosch et al., 1987a,b) do not accomplish this, usually describing expansion reasonably well but failing to simulate focal formation adequately.

The distinction between wavelike and non-wavelike epidemic spread has a strong theoretical basis, developed and described largely by Mollison (1972, 1977, 1978). The application of this theory to phytopathological problems requires a suitable model of disease spread and a mathematical description of dispersal that is both simple and realistic. The aim of this chapter is to discuss a few of the approaches that may be taken to model spread and to show how dispersal can be described in a way that leads naturally to the production of disease foci.

DIFFUSION MODELS

Models of population spread have traditionally been derived from an analogy to molecular diffusion, in which spread occurs as a series of random, infinitesimal jumps (e.g., Fisher, 1937; Skellam, 1951). The starting point for the analysis is Fick's law: The rate of transport of material across a unit area is proportional to the local concentration gradient of the material. For one-dimensional diffusion,

$$J = -D\frac{\partial C}{\partial s}$$

where C is concentration ($M \cdot L^{-1}$) at location s, J is flux ($M \cdot T^{-1}$) of material along the s-axis at location s, and D is a proportionality constant known as the diffusivity ($L^2 \cdot T^{-1}$). If J is constant with respect to s, as much material flows into any location as flows out and the concentration does not change. If a flux gradient exists at s, however, the concentration at s will change, at a rate equal to the value of the flux gradient:

$$\begin{aligned}\frac{\partial C}{\partial t} &= \frac{-\partial J}{\partial s}\\ &= \frac{\partial(D\, \partial C/\partial s)}{\partial s}\\ &= D\frac{\partial^2 C}{\partial s^2}\end{aligned}$$

where t is time. (The last equality is true only if D is constant with respect to s.) This is Fick's equation of diffusion, which relates the rate of change of concentration to its spatial distribution at any instant. By replacing C with population size, the same equation can be applied to population spread, predicting (for example) that a spreading population will eventually be normally distributed in space, with a variance equal to $2Dt$.

The basic diffusion model is of limited usefulness as it stands, because few real populations move completely at random. The model has great flexibility, however, and has been extended to cover a wide variety of situations (Okubo, 1980). Combining diffusion with directional spread (drift) leads to the Fokker–Planck equation, which forms the basis of Gaussian plume models of spore dispersal (Aylor, 1978). Combining it with simple population growth models produces what are sometimes called reaction-diffusion models, allowing the interaction of spread and reproduction to be investigated. For example, a logistic growth term can be added, resulting in an equation investigated many years ago by Fisher (1937):

$$\frac{\partial y}{\partial t} = D\frac{\partial^2 y}{\partial s^2} + ry(1 - y) \tag{1}$$

where y is population size at location s and time t as a proportion of the carrying capacity, and r is the intrinsic rate of population increase (to phytopathologists, the apparent infection rate). Equation (1) was shown by Fisher to predict wavelike population spread, in which the population gradient at any time is simply a translation of the gradient at some earlier time, the speed of translation being constant.

Diffusion models have a number of shortcomings, some of which are discussed from a mathematical viewpoint by Mollison (1972). He points out that the wavelike spread predicted by reaction-diffusion models is, on the basis of a more general analysis, expected to occur only under certain rather restrictive conditions, which are discussed in detail later in the chapter. This suggests that such models may give misleading indications about the behavior of spreading populations.

The most important consideration for phytopathologists is whether models based on the diffusion analogy can produce the kinds of spatial behavior that are known to occur during plant disease epidemics. Spread simulated by these methods tends to be very regular in form, a consequence of the fact that spread occurs by infinitesimal jumps; the large number of such jumps averages out any irregularities that may develop. One of the principal features of the spread of plant pathogens, the formation of disease foci, does not occur in diffusion models, because the rate and direction of diffusive spread depend only on the local population gradient. This means that disease increase at any point cannot be influenced by distant sources of infection, so discontinuities in disease occurrence cannot arise. For pathogens that are dispersed only very short distances the diffusion analogy may be acceptable, although it needs to be applied with care (see Minogue, 1986). In general, however, the inability of such models to simulate the formation, enlargement, and coalescence of foci makes them poorly suited to investigating the problem of disease spread.

SPATIAL CONTACT MODELS

Deterministic Models

Mollison (1972) has demonstrated that diffusion models tell only part of the story about population spread. He has also made major contributions to the development of less restricted models that exhibit behavior both richer and more realistic than diffusion models. His spatial contact models explicitly allow 'communication' (in the present context in the form of spore dispersal) between spatially separated locations, and the way in which this communication occurs (i.e., its mathematical description) can be varied to produce a wide range of spatial behaviors.

It is convenient to begin the discussion of these models by considering how the familiar logistic equation can be broadened to include spatial effects. The logistic model is one of a general class of epidemic models that assume the rate of increase of infected individuals (Y) in a population of total size N to be proportional to the product of the number of susceptible individuals remaining (''susceptibles'' S) and the number of infectious individuals present (''infectives'' I):

$$\frac{dY}{dt} = cSI \tag{2}$$

where c is a proportionality constant. The logistic equation results from the assumption that ''infected'' and ''infectious'' are equivalent, so that $I = Y$ and $S = (N - Y)$. Thus

$$\frac{dY}{dt} = c(N - Y)Y \tag{3}$$

In this equation, Y is implicitly a function of time. By considering Y also to be a function of space, the equation becomes a spatial model, but an incomplete one: It contains no mechanism by which the pathogen can be transmitted from one location to another. Including dispersal in the model remedies this deficiency, by allowing the rate of infection of individuals at location s to depend on all the infectives in the neighborhood of s, not just those exactly at s. The idea of the neighborhood of s, however, needs to be made more precise.

To do this, Mollison (1972, 1977, 1978) introduces the notion of a contact distribution, which in phytopathological context I have preferred to call a dispersal function (Minogue, 1986). This is simply the probability density function for the distance a spore travels from its source (s_o) to its point of deposition (s). For an epidemic occurring in one dimension, I will denote this function $f(x)$, where $x = s - s_o$. Since the dispersal function measures the relative contribution of infectives at s_o to the infection of susceptibles at s, it can be used as a weighting function to average the contributions of all infectives around s. Let the number of infectives at s_0 at time t be $Y(s - x, t)$; let these infectives produce spores at an average rate R per infective per unit time; and let $f(x)\triangle x$ be the probability that a spore produced at s_o will be deposited in the small interval between s and $s + \triangle x$. Then the number of spores produced at s_o that are deposited in that interval is

$$\Delta n(s, x, t) = RY(s - x, t)f(x)\ \Delta x$$

per unit time. The total deposition rate is found by letting the interval $\triangle x$ approach zero and integrating over all distances x about s:

$$n(s, t) = \int_{-\infty}^{+\infty} RY(s - x, t)f(x)\, dx$$

Thus, $n(s, t)$ is the total spore production by all infectives, per unit time, weighted by the probability the spores will reach s. Dividing by R (the rate of spore production per infective) give us the effective number of infectives at s. That is, the spore deposition rate at s, and thus the rate of infection of susceptibles, is the same as if there were $n(s, t)/R$ infectives concentrated at s and no dispersal between locations was occurring. This term can then replace the quantity I in equation (2), giving

$$\frac{\partial Y(s, t)}{\partial t} = c(N - Y(s, t)) \int_{-\infty}^{+\infty} Y(s - x, t)f(x)\, dx \tag{4}$$

as the spatial analogue of equation (3).

The chief benefit of the model in this form (which can easily be extended to two spatial dimensions) is that the explicit inclusion of a general dispersal function $f(x)$ allows the spatial development of the epidemic to be related to the properties of that function. One of Mollison's most important accomplishments has been to show that the regular, wavelike, constant-velocity spread characteristic of diffusion models will occur if and only if the integral

$$\psi(g) = \int_{-\infty}^{+\infty} f(x) \exp(gx)\, dx \tag{5}$$

converges for some real number $g > 0$ (Mollison, 1972, Theorem 2). The integral will converge only if the density $f(x)$ declines rapidly enough with x—that is, more rapidly than the exponential function is increasing. If not, the relatively large probability of long-distance dispersal results in spread characterized by neither a clearly defined waveform nor a clearly defined velocity—the disease gradient becomes progressively flatter, and the velocity of spread progressively faster, as the epidemic proceeds. The importance of Mollison's discovery cannot be overstated: once a plausible form of the dispersal function has been derived from measurements of the dispersal gradient or from theoretical considerations, the convergence or divergence of the integral in equation (5) provides immediate and far-reaching insight into how the spreading epidemic can be expected to behave. The form of the equation is particularly convenient for this purpose, since it is just the definition of the moment-generating function associated with $f(x)$. Such functions have been extensively tabulated (e.g., Oberhettinger, 1973), so that the existence of $\psi(g)$ can often be determined simply by looking up the appropriate entry in a table.

Stochastic Models

The model expressed in equation (4) is a deterministic model, referring (at best—see Mollison, 1977) to average behavior over many possible epidemics. Stochastic elements will, under most circumstances, cause individual epidemics to deviate markedly from the average, notably by the formation of spatial discontinuities in population density (i.e., disease foci). To investigate fully the variety of behavior that individual epidemics can exhibit, it is necessary explicitly to include stochastic elements in the model. In doing so, analysis is necessarily replaced largely by a simulation protocol (Mollison, 1972; Minogue and Fry, 1983) which follows each individual spore during dispersal and reproduction. The distance that a given spore travels is chosen at random in such a way that the spore population as a whole obeys a particular dispersal function. Over many infection cycles, large-scale patterns develop in the population distribution, whose properties reflect the properties of the dispersal function. Some of those properties are described below for two of the most important dispersal functions.

Dispersal Function

Exponential distribution. Probably the simplest assumption about spore dispersal is that the crop acts as a filter through which the spore passes, and that the spore has a constant probability of deposition per unit of distance traveled. That is, the probability of deposition in a small distance increment Δx is the same regardless of how far the spore has already traveled. If this probability is $k\ \Delta x$ and is the same for all spores, then by standard methods it can be shown that the proportion of spores deposited within a distance x of their source is

$$F(x) = 1 - \exp(-kx)$$

The dispersal function is the probability density,

$$f(x) = \frac{dF(x)}{dx}$$
$$= k \exp(-kx)$$

This is an exponential distribution, and it underlies the so-called Kiyosawa and Shiyomi (1972) equation for dispersal gradients:

$$m(x) = a \exp(-bx)$$

$m(x)$ being the number of spores deposited a distance x from the source. The constant a depends on source strength, and b is the gradient parameter (see Minogue, 1986).

The moment generating function of the exponential distribution is

$$\psi(g) = \int_0^\infty k \exp(-kx) \exp(gx)\, dx$$
$$= \frac{k}{k - g}$$

for $g < k$. (For simplicity of presentation I am considering only spread in the positive direction in a one-dimensional population, so the lower limit of integration is zero instead of $-\infty$.) Since $\psi(g)$ exists, the spread of a population that disperses according to an exponential dispersal function is expected to be wavelike in character and to occur at a constant velocity. Simulation supports this expectation in one-dimensional epidemics (Mollison, 1972; Minogue and Fry, 1983). In addition, it seems intuitive that the rate of spread of an epidemic should be determined largely by the variance of the dispersal function, a spread-out dispersal function producing a fast-spreading epidemic. Mollison's analytical results and simulation both support this intuition. Spread in two dimensions has been studied little so far, but preliminary indications (unpublished) are that the velocity of spread for an exponentially dispersing population is constant and depends on the variance of the dispersal function in a manner similar to spread in one dimension.

Pareto distribution. A useful probability function for simulating and investigating non-wavelike spread can be developed from a modification of Gregory's (1968) gradient equation,

$$m(x) = ax^{-b}$$

where $m(x)$, as before, is number of spores trapped or infections counted at a distance x from the source, and a and b are positive parameters (unrelated to the a and b of Kiyosawa and Shiyomi's equation). Although useful for many purposes, Gregory's equation suffers from a number of shortcomings (Minogue, 1986), among them the fact that $m(x)$ must approach infinity near the source ($x = 0$), which clearly is biologically impossible. The practical result is that the equation usually overpredicts deposition or infection near the source (McCartney and Bainbridge, 1984). To overcome this, Mundt and Leonard (1985) introduced another parameter to Gregory's equation,

$$m(x) = a(x + c)^{-b} \qquad (6)$$

which (if $c > 0$) allows $m(x)$ to be finite at $x = 0$. Mundt and Leonard (1985) have shown that this modified equation provides a good fit to cereal and bean rust data. Relative to the exponential function, equation (6) predicts more deposition near the source and far from the source, and less deposition at intermediate distances (Fig. 1).

The functional form of equation (6) is similar to the well-known Pareto probability distribution,

$$f(x) = \alpha\theta^{\alpha} (x + \theta)^{-(\alpha + 1)}$$

where α and θ are positive parameters. A spore population dispersing according to a Pareto distribution will produce a dispersal gradient obeying equation (6), with $c = \theta$, $b = (\alpha + 1)$, and a dependent on source strength. This relationship, together with the demonstrated applicability of equation (6) to field data, suggests that the Pareto distribution may be a useful model for spore dispersal.

An heuristic rationale for the Pareto distribution as a spore dispersal function may be suggested as follows. Recall that a negative exponential dispersal gradient results if one assumes that the probability of a spore depositing on foliage in a unit distance remains constant with distance from the source, and that this probability is the same for all spores. In fact, due to variation in spore size and density and the path taken during transport (through the crop or over it), spores differ in their probability of deposition: larger spores, for example, will tend to deposit close to the source, and smaller spores farther away. Variability in spore size means that relative to a uniform spore population, there are more large and small spores and fewer of intermediate size, so

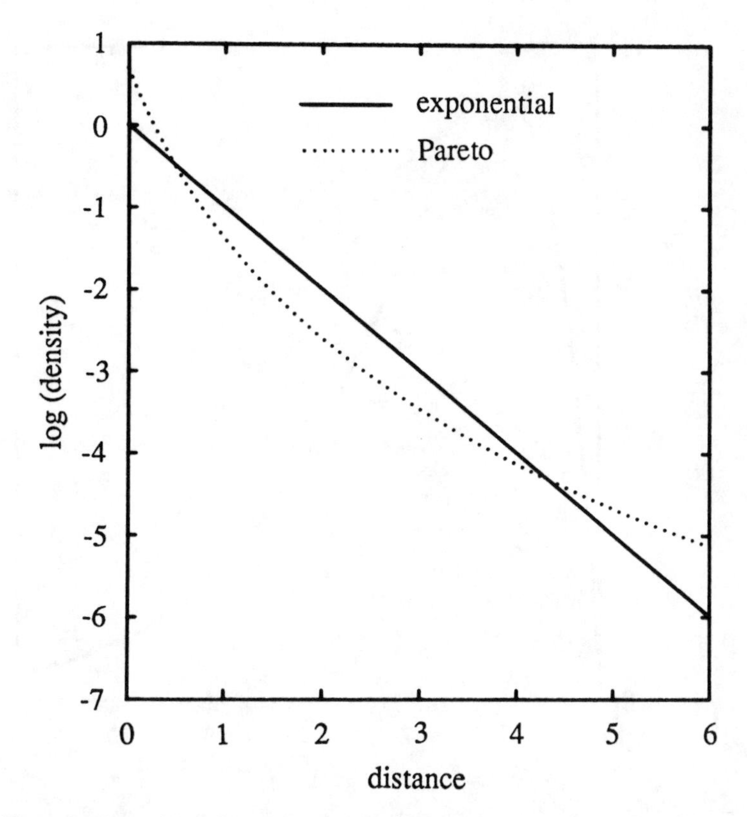

Figure 1 Natural logarithm of probability density as a function of distance from the source, for an exponential distribution and a Pareto distribution. Both distributions have a mean of 1; the Pareto distribution has $\alpha = 2$.

one can expect more deposition at near and far distances, and less in the middle. This is exactly the situation addressed by equation (6) and the Pareto distribution.

By standard techniques for deriving contagious distributions (see, for example, Mood et al., 1974, p. 124 for a brief discussion), it can be shown that the Pareto distribution arises if we begin with the same assumptions that led to the exponential dispersal function, except that the k parameter (probability of deposition per unit distance) is allowed to be a random variable with a different value for each spore. If that variable has a gamma probability distribution with scale parameter θ and shape parameter α, the distance traveled by a spore before deposition will have a Pareto distribution. That the probability of deposition for real spore populations does, in fact, obey a gamma distribution is at the moment entirely speculative, but the distribution is at least plausible and flexible enough to use as a tool for developing concepts. The gamma distribution associated with the Pareto distribution of Fig. 1 is shown in Fig. 2.

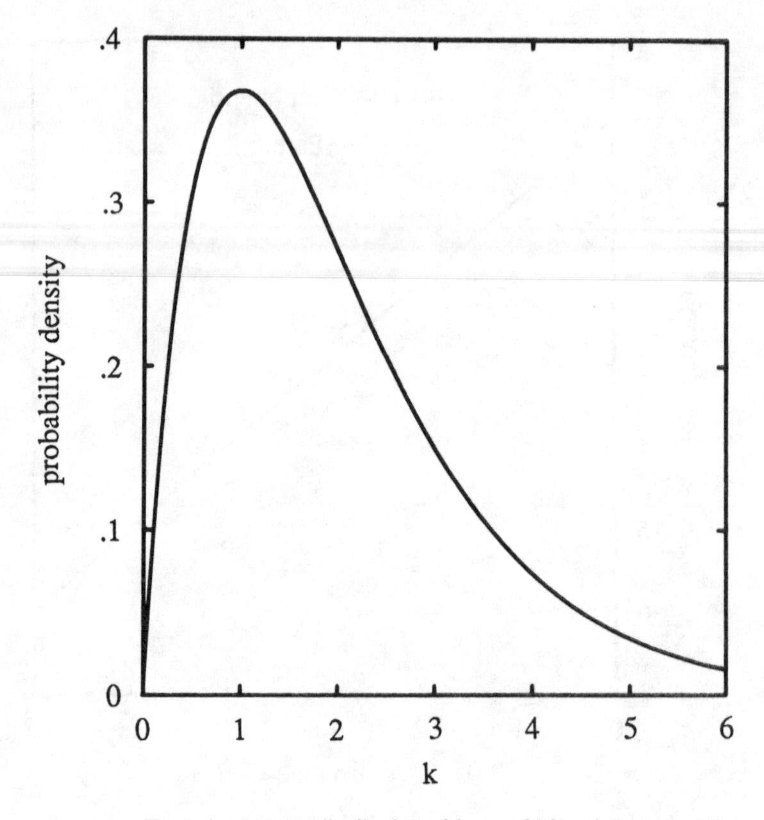

Figure 2 Gamma distribution with $\alpha = 2, \theta = 1$.

The Pareto distribution is particularly interesting for our purposes because it does not have a moment generating function $[\psi(g)]$, and thus by Mollison's criterion is not expected to produce wavelike epidemics. The mean and variance of the distribution may be finite or infinite, depending on the value of α:

$$\mu = \frac{\theta}{\alpha - 1} \qquad \text{for } \alpha > 1$$

$$\sigma^2 = \frac{\alpha\,\theta^2}{(\alpha - 1)^2(\alpha - 2)} \qquad \text{for } \alpha > 2$$

The mean is infinite for $0 < \alpha \leq 1$, and the variance is infinite for $0 < \alpha \leq 2$. A dispersal function with such properties is bound to produce interesting simulations, if nothing else; as will be seen shortly, the simulations are also impressively realistic.

The concept of infinity in this context may perhaps need some explanation. Clearly, the distance of dispersal for a finite population of spores on a finite planet can have neither an infinite mean nor an infinite variance, so the use of a probability function with these properties may seem unrealistic. In a practical sense, the result of a small α is that no matter how large the distance over which observations of spore dispersal are made, enough spores travel beyond that distance to make estimates of the mean or variance based on those observations unreliable. Figures 3 and 4 illustrate the expected value of mean and standard deviation estimates, respectively, for a range of Pareto distributions, when data collection is truncated at various distances from the source. In all cases θ is chosen so that 50% of the spores are deposited within 1 m of the source. When the mean or standard deviation exists, the expected values of the respective estimates eventually level off at the appropriate theoretical values. When these quantities do not exist, the estimates simply continue to increase without bound. It is suggested that a Pareto distribution with infinite variance (and possibly infinite mean) may be appropriate for data that show no convergence to constant estimates over practical scales of observation.

Because both the variance and the mean of the Pareto distribution may be infinite, it is sometimes necessary to use quantiles to compare it to other

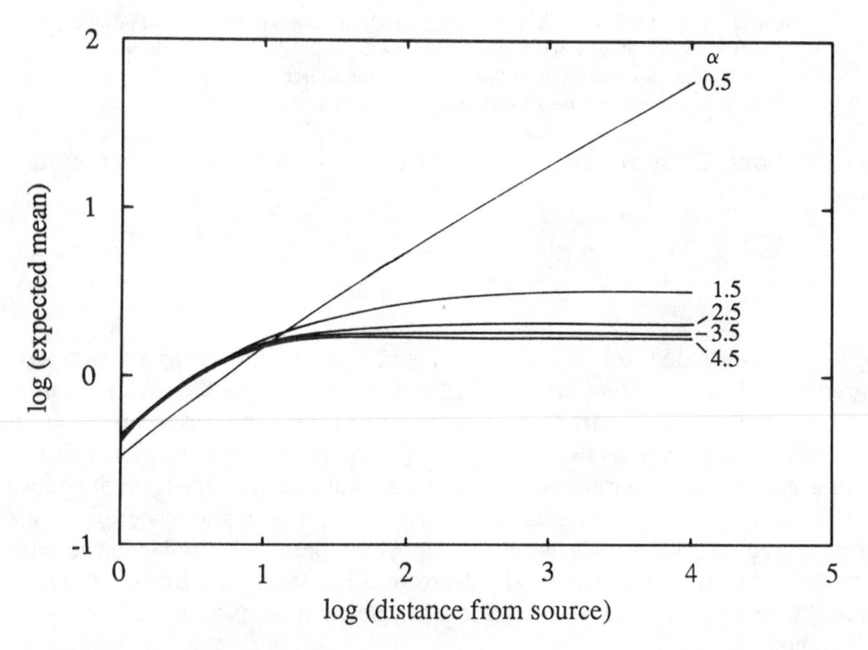

Figure 3 Expected value of the estimated mean distance of dispersal, for several values of α, when dispersal follows a Pareto distribution and sampling is truncated at various distances (meters) from the source. The median distance of dispersal in all cases is 1 m. Logarithms on both axes are to base 10.

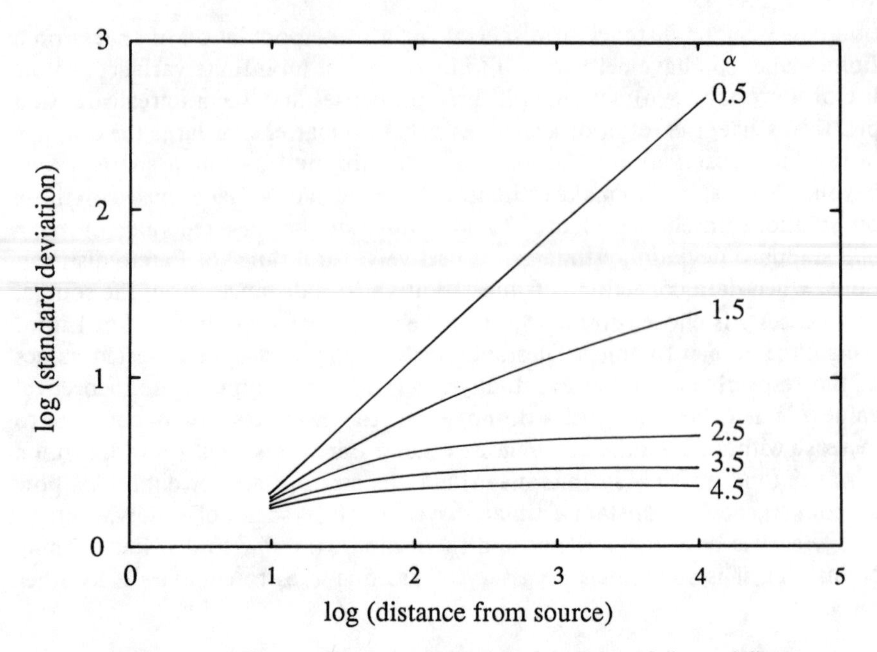

Figure 4 Expected value of the estimated standard deviation of dispersal distance, for several values of α, when dispersal follows a Pareto distribution and sampling is truncated at various distances (meters) from the source. The median distance of dispersal in all cases is 1 m. Logarithms on both axes are to base 10.

distributions. These are easily obtained from the cumulative distribution function

$$F(x) = 1 - \left(\frac{\theta}{x + \theta}\right)^{\alpha}$$

As mentioned, θ is a scale parameter, a large θ implying a short, steep gradient. However, comparison of equation (6) with the Pareto distribution shows that the counterpart of the b parameter (usually interpreted as the gradient steepness parameter) is not θ but α (i.e., $b = \alpha + 1$). In fact, neither θ nor α is a "gradient parameter" in the sense that I defined the term previously (Minogue, 1986); both interact to determine the shape and scale of the dispersal gradient. When α is large, the deposition pattern is similar to the exponential (Fig. 5); when it is small, deposition is concentrated near the source and far away, resulting in a gradient that defies a simple "steep" or "flat" terminology.

A consequence of this kind of dispersal pattern is that epidemics simulated with the Pareto distribution as the dispersal function exhibit features not seen when the "well-behaved" exponential distribution is used. One such feature is the production of distinct foci. Figure 6 shows a simulated epidemic

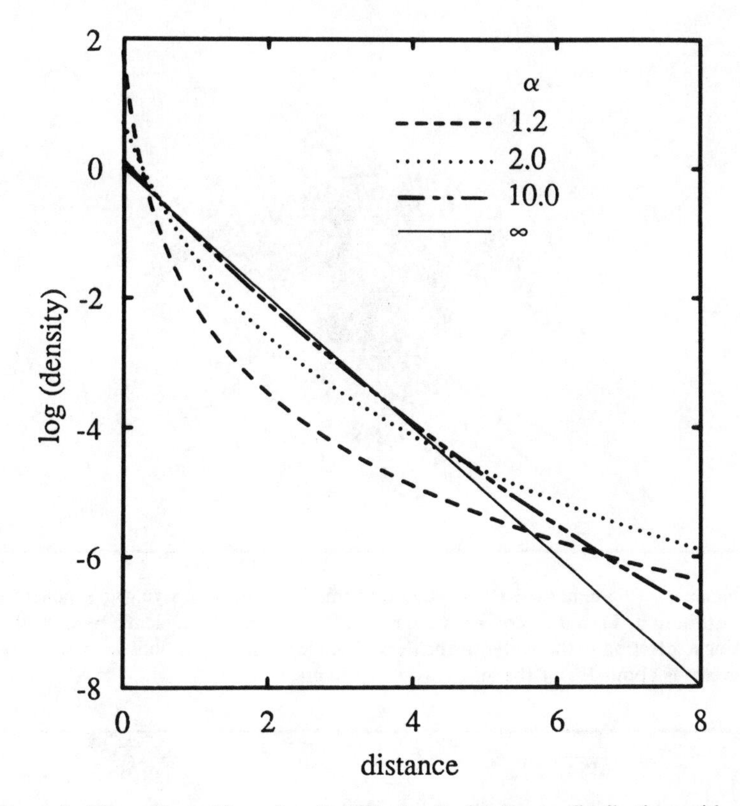

Figure 5 Natural logarithm of probability density for Pareto distributions with a mean of 1.

in which dispersal follows an exponential distribution, the mean distance of individual dispersal events being equal to about 3% of the long axis of the figure. The epidemic began with a single infection at the center of the figure. Although the distribution of infections is more scattered and irregular than would occur with most diffusion-based models, there is still a notable absence of satellite foci; only the main focus is present. In Fig. 7 the mean distance of dispersal is the same, but a Pareto distribution with $\alpha = 1.2$ is used to simulate spore dispersal. The main focus is denser and more compact, reflecting the higher probability of deposition near the source. In addition, however, clear satellite foci have developed, the result of the higher probability of longer-distance dispersal, relative to the exponential distribution. More precisely, the Pareto distribution produces foci because the probability of deposition per unit distance, after a steep initial decline, decreases very slowly with increasing distance from the source.

One way of characterizing the spread of an epidemic is to record the distance to the farthest infection (the front) as a function of time. For a well-behaved dispersal function like the exponential, this distance increases linearly

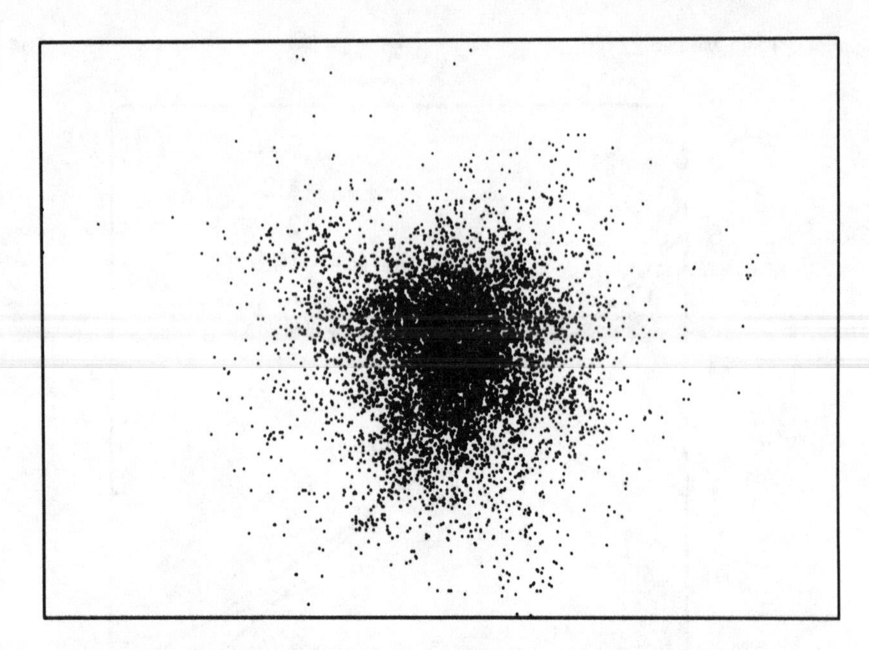

Figure 6 Two-dimensional simulated epidemic for dispersal obeying an exponential distribution. Each dot represents a simulated infection. The epidemic began with a single infection in the center of the field. The mean distance of individual dispersal events is about 3% of the long axis of the figure.

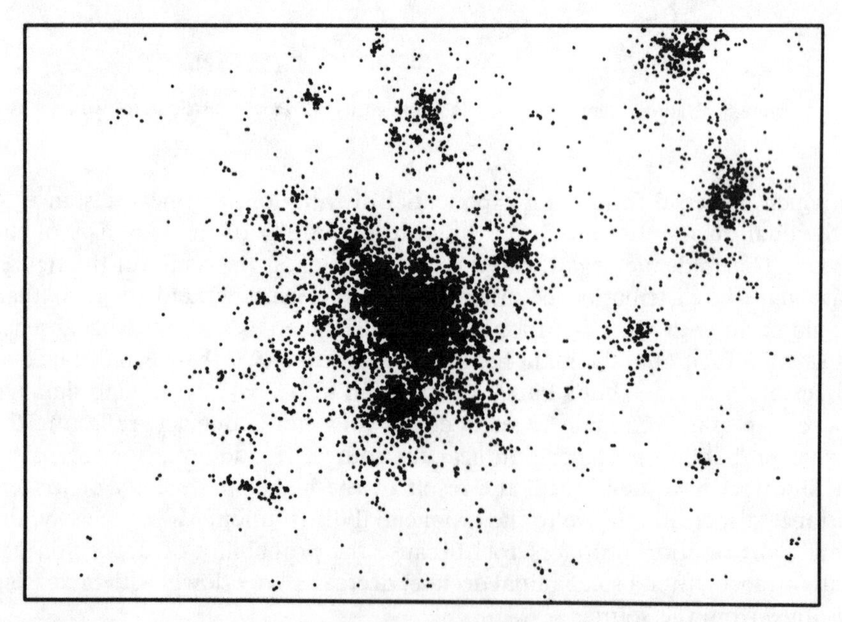

Figure 7 Two-dimensional simulated epidemic for dispersal obeying a Pareto distribution with $\alpha = 1.2$. Each dot represents a simulated infection. The epidemic began with a single infection in the center of the field. The mean distance of individual dispersal events is about 3% of the long axis of the figure.

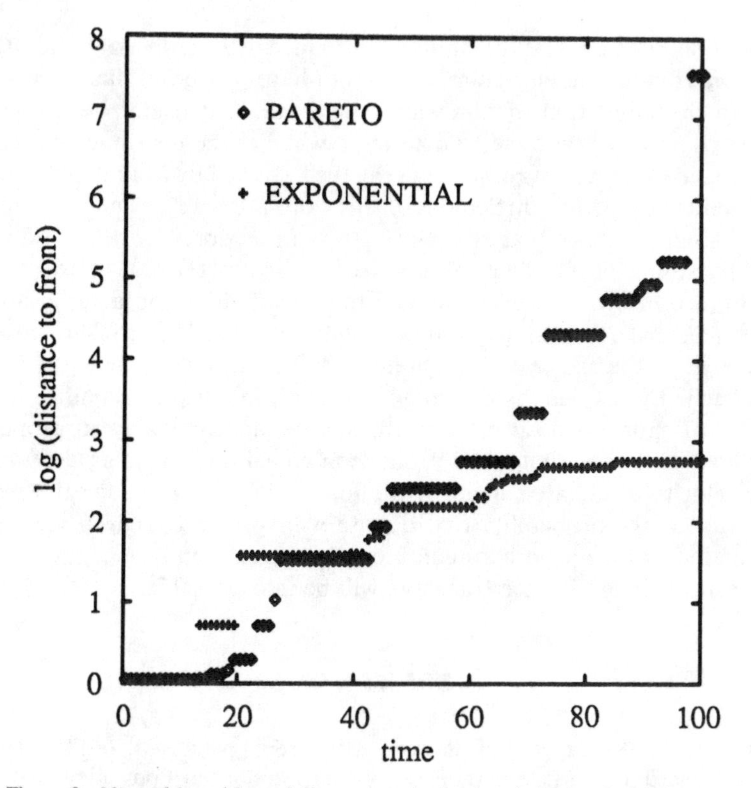

Figure 8 Natural logarithm of distance to the farthest infection from the source, for two one-dimensional simulated epidemics. Dispersal occurred according to either a Pareto distribution (α = 1.2) or an exponential distribution. The mean dispersal distance in both cases was 1 unit.

with time, the rate of increase—the velocity of spread—being proportional to the standard deviation of the dispersal function (Mollison 1972, 1977; Minogue and Fry, 1983). When the standard deviation does not exist, the front moves by "great leaps forward" (as Mollison called them), which may become increasingly large as the epidemic proceeds. Figure 8 shows the behavior of one-dimensional simulated epidemics using either a Pareto distribution (α = 1.2) or an exponential distribution to simulate spore dispersal. The mean distance of dispersal is the same (1 unit) in both cases. Even on a log scale the Pareto epidemic shows large jumps and an increasing velocity of spread.

CONCLUSIONS

The main intent of this chapter has been to illustrate and reinforce the idea that the spatial patterns produced by spreading pathogen populations are the direct result of the mechanisms by which those populations are dispersed, and that quite simple dispersal functions can produce realistic spatial patterns. The Pareto distribution, in particular, appears to have properties that make it well suited to the simulation of non-wavelike spread. A considerable amount of field work needs to be done to determine whether the distribution describes the dispersal of organisms other than the rusts studied by Mundt and Leonard (1985), and the practical difficulties of this work are not to be underestimated. Figures 3 and 4 indicate that even when 50% of the spores are deposited within 1 m of the source, it may be necessary to continue observations up to 100 times this distance in order to get reliable estimates of the mean and variance of dispersal distance, assuming that these quantities exist. The problem, however, does not lie in the choice of the Pareto distribution, but in the nature of dispersal itself. Even when, as is common among wind-dispersed pathogens, the majority of spores deposit close to the source, and only a small proportion travel long distances, that small proportion constitutes the main determinant of the velocity of disease spread. If we hope to be able to predict the velocity of spread, or the probability that disease will spread to a given region in a given period of time, or other quantities of practical significance, quantitative assessment of long-distance transport will be indispensable.

REFERENCES

AYLOR, D. E. 1978. Dispersal in time and space: aerial pathogens. *In* Plant Disease: An Advanced Treatise, Vol. II, How Disease Develops in Populations (ed. J. G. Horsfall and E. B. Cowling), pp. 159–180. Academic Press, Inc., New York.

BOSCH, F. VAN DEN, ZADOKS, J. C., and METZ, J. A. J. 1988a. Focus formation in plant disease. I. The constant rate of focus expansion. Phytopathology 78:54–58.

BOSCH, F. VAN DEN, ZADOKS, J. C., and METZ, J. A. J. 1988b. Focus formation in plant disease. II. Realistic parameter-sparse models. Phytopathology 78:59–64.

FISHER, R. A. 1937. The wave of advance of advantageous genes. Ann. Eugenics 7:355–369.

GREGORY, P. H. 1968. Interpreting plant disease dispersal gradients. Annu. Rev. Phytopathol. 6:189–212.

KIYOSAWA, S., and SHIYOMI, M. 1972. A theoretical evaluation of the effect of mixing resistant variety with susceptible variety for controlling plant diseases. Ann. Phytopathol. Soc. Jpn. 38:41–51.

MCCARTNEY, H. A., and BAINBRIDGE, A. 1984. Deposition gradients near to a point source in a barley crop. Phytopathol. Z. 109:219–236.

MCCARTNEY, H. A., and FITT, B. D. L. 1986. Spore dispersal in relation to epidemic models. *In* Plant Disease Epidemiology, Vol. I: Population Dynamics and Management (ed. K. J. Leonard and W. E. Fry), pp. 311–345. Macmillan Publishing Company, New York.

MINOGUE, K. P. 1986. Disease gradients and the spread of disease. *In* Plant Disease Epidemiology, Vol. I, Population Dynamics and Management (ed. K. J. Leonard and W. E. Fry), pp. 285–310. Macmillan Publishing Company, New York.

MINOGUE, K. P., and FRY, W. E. 1983. Models for the spread of disease: model description. Phytopathology 73:1168–1173.

MOLLISON, D. 1972. The rate of spatial propagation of simple epidemics. Proc. 6th Berkeley Symp. Math. Stat. Prob., Vol. 3. University of California Press, Berkeley, Calif., pp. 579–614.

MOLLISON, D. 1977. Spatial contact models for ecological and epidemic spread. J. R. Stat. Soc. Ser. B 39:283–326.

MOLLISON, D. 1978. Markovian contact processes. Adv. Appl. Probab. 10:85–108.

MOOD, A. M., GRAYBILL, F. A., and BOES, D. C. 1974. Introduction to the Theory of Statistics, 3rd ed. McGraw-Hill Book Company, New York.

MUNDT, C. C., and LEONARD, K. J. 1985. A modification of Gregory's model for describing plant disease gradients. Phytopathology 75:930–935.

OBERHETTINGER, F. 1973. Fourier Transforms of Distributions and their Inverses: A Collection of Tables. Academic Press, Inc., New York.

OKUBO, A. 1980. Diffusion and Ecological Problems: Mathematical Models. Springer-Verlag New York, Inc., New York.

SKELLAM, J. G. 1951. Random dispersal in theoretical populations. Biometrika 38:196–218.

7

Spatial Heterogeneity and Disease in Natural Populations

Helen Miller Alexander

INTRODUCTION

In recent years, many plant pathologists have been interested in the dynamics of plant–pathogen interactions in natural populations as sources of new approaches or strategies for disease control in agriculture (e.g., Browning, 1974, 1980; Schmidt, 1978; Segal et al., 1980; Dinoor and Eshed, 1984). Interest in this area was intensified in the years following the 1970 epidemics of southern corn leaf blight in cytoplasmic male-sterile maize, when attention was focused on the extreme genetic and cultural uniformity of the major crop species in Western agriculture (Anonymous, 1972; Ullstrup, 1972). In contrast, it appeared that disease epidemics were rare or of short duration in natural plant populations and it was concluded that plants and pathogens in nature were in ecological and evolutionary balance. This state was largely attributed to the ecological and genetic diversity of natural populations and communities (e.g., Browning, 1974, 1980; Dinus, 1974; Schmidt, 1978; Segal et al., 1980; Dinoor and Eshed, 1984). By mimicking aspects of these natural populations and communities, it was postulated that disease control in agriculture could be achieved

Note: Department of Biology, University of Louisville, Louisville, KY 40292, US. Current address: Departments of Botany and Systematics & Ecology, University of Kansas, Lawrence, KS 66045, US.

in more natural, long-lasting ways. Thus disease control strategies such as multilines and cultivar mixtures and other methods of reduction of crop uniformity were partly justified as ways of incorporating lessons learned from studies of natural populations into modern agriculture (e.g., Browning, 1974, 1980; Burdon, 1978; Schmidt, 1978; Burdon and Shattock, 1980).

A detailed examination of the data available on plant–pathogen interactions in natural populations shows, however, that critical experiments have not been performed to test which factors determine the patterns of disease incidence in nature and the dynamics of interactions between plants and their pathogens. This lack of information is due largely to two reasons. First, plant ecologists and evolutionary biologists, who traditionally explore the factors that control the numbers, spatial distributions, and ecological and genetic interactions of organisms in natural populations, have historically ignored pathogens of plants (Harper, 1977). Factors such as the scientists' deficiencies in microbial techniques and the low visibility of many diseases in nature were probably responsible for this lack of interest in natural plant–pathogen interactions. Second, although plant pathologists have studied disease in natural populations, their work is generally geared toward specific agricultural objectives instead of understanding the general dynamics of disease processes in natural populations. For example, natural populations of plants and their pathogens have been examined to identify useful resistance genes, determine the feasibility of using pathogens for biocontrol of weedy plants, and identify alternative hosts or agriculturally important pathogens (e.g., Dinoor and Eshed, 1984). Although the agricultural objectives were fulfilled, the data have not always been collected in a form that can be directly used to understand ecological and genetic processes.

In this chapter I first describe the major ecological and genetic differences between natural and agricultural populations. I then examine three aspects of spatial heterogeneity (variation in plant density, plant genotypic composition, and plant location with respect to the physical environment) that are likely to affect disease in nature. For each, I first state its predicted theoretical impact on disease epidemiology and the expected ecological and genetic consequences for the plant population. The consequences for the plant population in each case will depend on the assumption that infection by a pathogen will be deleterious to an individual plant; specifically, that it will reduce the plant's probability of survival and/or reproductive output and thus reduce its fitness. The large number of studies on effects of disease on plant yield support this assertion, since most yield components measured on the individual crop plant can be equated with fitness components of individuals in natural populations (Antonovics and Alexander, 1988). Only recently, however, has research been focused on fitness effects of disease in natural populations (e.g., Burdon and Shattock, 1980; Augspurger, 1983; Alexander and Burdon, 1984; Clay, 1984; Parker, 1986; Antonovics and Alexander, 1988). I then test the theoretical prediction of the effect of heterogeneity on disease epidemiology by examining the results of previous studies. In presenting the literature, my intention is to include studies by scientists outside the discipline of plant pathology. My goal is not an exhaustive review of the literature, but a discussion of published research that illustrates what is and is not known. Finally, I explore the need of particular types of studies to increase our knowledge of disease in natural systems.

DIFFERENCES BETWEEN NATURAL AND AGRICULTURAL PLANT POPULATIONS

The major difference between natural and agricultural plant populations is that in the former, there is greater variability in the plant population itself and in its biotic and abiotic environment (Table 1); much of this variability has a spatial component. Comparisons of plant community composition reveal that agricultural fields are planted to one species of plant while natural communities are usually composed of multiple species. Within a plant population, genetic diversity is much lower in most agricultural systems, due to the planting of a single crop variety. The phenotypes of plants in agricultural populations are usually less diverse than in natural populations: plants are often of similar size, age, and developmental stage due to a single planting date and breeding for cultivars that mature synchronously. Crop plants are also usually planted in rows of regular spacing, as opposed to the random or clumped spatial patterns more common in natural populations. Finally, the physical environment in which agricultural plants live is less variable: soil may be tilled, treated with fertilizer, and irrigated in an attempt to provide a uniform habitat for plant

TABLE 1 Comparison of Agricultural and Natural Plant Populations for Attributes That May Affect Disease Epidemiology[a]

	Agricultural	Natural
Plant community		
Species composition	One to few species	Many species
Plant population		
Genotypic composition	One to few genotypes (cultivars)	Many genotypes
Age, developmental stage	Uniform	Variable
Plant spacing	Uniform	Variable
Physical environment	Uniform	Variable
Factors determining community and population characteristics	Humans, abiotic and biotic environment	Abiotic and biotic environment

[a]Extreme differences are given, with the realization that a continuum of types of communities and populations exist.

growth. Clearly, most agricultural fields are not completely uniform for these traits: weeds create species diversity, off-type plants give within-species genetic diversity, and variation in topography can result in areas of better and poorer plant growth. However, in comparison to natural populations, whether in a forest, wetland, or grassland community, large differences in these categories exist.

Besides the categories above, another important difference between natural and agricultural systems is the type of factors that determine the dynamics of the plant population. In agricultural populations, humans control many plant population characteristics by actively choosing the crop species, cultivar, planting date and density, row spacing, and methods of soil preparation and cultivation. In contrast, the characteristics of natural plant populations develop as a result of interactions of biotic and abiotic factors in the ecosystem, including the process of evolution. This is true whether one is considering a plant growing in an undisturbed tropical forest or along a weedy roadside. In the latter case, human actions (e.g., roadside mowing, lead deposits from passing cars) clearly affect plants, but the effect is distinct from agricultural situations since humans are largely changing broad environmental factors that may affect plant survival and reproduction rather than actively choosing specific plant traits and genotypes. The differences between natural and agricultural systems obviously form a continuum; for example, many forage and tree crops are intermediate in regard to the variability in the plant population and its environment and in terms of the effect of humans in determining these characteristics. Natural plant populations also differ in these attributes: some natural communities are composed of single species and the extent of genetic variability present in a population is highly dependent on its breeding system and the extent of vegetative reproduction. However, in general, the contrasts presented are both representative and important because of their effect on epidemiological processes.

ASPECTS OF HETEROGENEITY AFFECTING DISEASE IN NATURE

Plant Density

Prediction. *A positive association exists between plant density and disease incidence and disease spread.* At high plant densities, disproportionately high disease levels are expected since both more target plants are available to the pathogen and because short interplant distances facilitate disease spread when propagule dispersal gradients are steep. This positive relationship between plant density and disease incidence and spread is the most common pattern in agricultural plant populations, accounting for 62% of the host–pathogen combinations reviewed by Burdon and Chilvers (1982). Negative correlations between host density and disease incidence are, however, also found [(27% of the host-pathogen combinations analyzed by Burdon and Chilvers (1982)]. These negative relationships can occur due to indirect effects of host density on plant growth, microenvironment, vector behavior, or plant community composition that may also alter disease dynamics. Further, factors that affect the relative importance of within- versus between-plant transmission of disease (e.g., plant size, pathogen dispersal gradient) will also determine how dependent disease processes are on host density (Burdon and Chilvers, 1982).

Potential consequences. A positive relationship between host density and disease spread could help address a dominant ecological topic, that of regulation of population size (Antonovics and Levin, 1980). Although populations have the potential for exponential growth, dramatic increases in population size are not common. Density-dependent processes, where plant survivorship or reproduction decrease as plant density increase, are likely to be important in maintaining a relatively constant population size over time. In the case of disease processes, the following scenario could occur. When a plant population is first established in a new area, plant density is low. The probability of plants within the population becoming diseased is also low, both because the probability of a host-specific pathogen encountering a rare plant is low and because, even if this encounter does occur, the newly diseased plant is often spatially isolated from its conspecifics. As the plant population increases in size within a given area, however, disease dynamics change. Pathogen propagules are likely to encounter a common plant, the amount of pathogen inoculum increases with each successful infection, and disease spread between plants becomes increasingly more likely as interplant distances decrease. The high levels of disease present at high host densities could severely affect the fitness of individual plants. This impact of disease at high densities may also be intensified by increased intraspecific competition between plants at high densities, reducing the vigor and reproductive output of diseased plants. As plant densities decline due to pathogen-induced death or lowered seed production, rates of disease spread would also be expected to decline as the number of hosts is diminished and the distances separating susceptible plants are greater. Thus the size of both plant and pathogen populations could be regulated by a negative feedback system controlled by density-dependent processes (Burdon and Shattock, 1980; Burdon and Chilvers, 1982).

The same process could also be important on the community level as one of the factors that explains how many plant species exist in the same community. Janzen's (1970) model on the importance of density-dependent responses of species-specific herbivores in the maintenance of a species-rich plant community can also be applied to plant–pathogen interactions. Most seeds produced by a plant fall close by, leading to a density gradient with high seed and seedling densities near the parent plant and low seed and seedling densities farther from the source. If host-specific pathogens cause a greater proportion of plants to be diseased at high plant densities, the probability of seedling survival will be greatest at distances far from the parent where seed densities are lowest. Due to this density effect, each plant species density may be maintained at a level lower than the number predicted simply by the abiotic resources of the environment. Coexistence of multiple species would then be possible: an individual of, for example, species A has a greater probability of survival close to an individual of species B than would a conspecific. Other factors, such as changes in interspecific plant competition in the presence and absence of disease, have also been postulated to be important in maintaining high species diversity in plant communities (Burdon and Shattock, 1980; Burdon, 1982). A simple model that integrated both density-dependent disease spread and competition between plants of different species led to the maintenance of two plant species when each was exposed to host-specific pathogens (Chilvers and Brittain, 1972).

Evidence. Despite both the general prediction of a positive relationship between plant density and disease spread, and its important implications at the plant population and community level, there have been few studies of this topic in nonagricultural populations. In a study of an inflorescence disease of *Plantago lanceolata* caused by *Fusarium moniliforme* var. *subglutinans,* the incidence of disease was analyzed in relation to the density of inflorescences on three spatial scales (25-, 6.25-, and 1-m^2 quadrats) and to the density of flowering plants on one scale (1-m^2 quadrats) (Alexander, 1984). In all cases, areas with low densities of inflorescences or flowering plants had a lower probability of containing at least one diseased inflorescence than expected if patterns of disease incidence were independent of density. These results suggest a positive relationship between host density and the presence of disease, which could occur if the pathogen is maintained above a threshold host density. In those areas where the disease was found, however, there was a negative or lack of density dependence between the number diseased and host density. The pathogen was apparently unable to exploit high host densities, probably because of the long disease cycle and the reduced efficiency of disease spread in the heterogeneous environment of a mixed-species field.

Descriptive studies, such as Alexander (1984), are less powerful than experimental studies that manipulate host density and analyze the effect on disease incidence or spread. A study that combines both descriptive and experimental methods concerns damping-off diseases of seedlings of the tropical tree, *Platypodium elegans* (Augspurger, 1983; Augspurger and Kelly, 1984). In a descriptive study (Augspurger, 1983), the locations of individual seedlings beneath isolated parent trees were mapped so that seedling densities and distances from the parent tree were known. The fate of individual seedlings was determined by following them for one year past germination. Highest incidence of disease caused by damping-off fungi (unknown species) occurred in the regions with highest seedling density, which were closest to the parent tree. The descriptive nature of the study meant, however, that other factors, such as distance from parent tree, were confounded with plant density. Thus an experimental manipulation was done in which seeds were placed out at high or low densities in quadrats cleared of naturally occurring seed, at three distances from the parent tree (Augspurger and Kelly, 1984). Again, the fates of individual seedlings were monitored. In this experiment, the effect of density on plant disease levels was found to be dependent on the distance of the seedling from the parent tree. Directly under the tree, both high- and low-density treatments had equally high levels of disease, presumably because of a favorable environment and high inoculum levels resulting from many years of high seedling numbers and high disease levels. Far from the tree, however, the high-density treatment had significantly higher disease levels than the low-density treatment. The inoculum level was lower at this far distance and thus the disease level was dependent on the rate of disease spread between plants, a process that occurs faster at high densities.

Augspurger's work provides experimental evidence for Janzen's (1970) model postulating that seed dispersal may function to remove progeny from areas of high density and high mortality near the parent tree. Such patterns of recruitment, if true for many species, would suggest that pathogens may act to promote the maintenance of high plant species diversity. The general nature of her results with *Platypodium elegans* is suggested by her survey of seed dispersal and seedling fates of nine tropical trees (only one tree per species was studied) (Augspurger, 1984). Pathogens accounted for the largest proportion of all seedling deaths in six of the nine species, and dispersal away from the parent tree reduced overall mortality in eight species in at least one of the time intervals studied. The exact cause of the decrease in mortality (i.e., distance from tree or seedling density), however, cannot be separated in this study.

Another source of data relating host density and disease levels comes from planned introductions of pathogens to weedy populations of their host plants in biocontrol projects (Burdon and Chilvers, 1982). In the case of skeleton weed, *Chondrilla juncea,* in Australia, a rust pathogen *(Puccinia chondrillina)* of plants in their native range in Europe proved to be an effective biocontrol agent. In the years following introduction of *P. chondrillina* to Australia, plant densities decreased dramatically, to densities more typical of the plant in Europe (Cullen and Groves, 1977). These results provide suggestive data that in natural populations of skeleton weed, disease may be one factor that regulates population size.

These few studies, showing that disease spread was positively density dependent in nature, provide support for the concept that disease is capable of regulating host population size and affecting community species diversity. Many questions, however, remain unanswered, such as the spatial scales over which density dependence acts and the effects of clumped spatial patterns on disease levels. Answers to these questions will come only from experimental studies that manipulate densities in natural populations.

Genetic Composition of the Plant Population

Prediction. *Disease incidence and spread are reduced in genetically variable plant populations.* Based on the experience of extensive plant breeding programs, we can expect that genetic variation occurs among plants in nature in their susceptibility to disease. Thus a natural population of plants might be predicted to be a mixture of individuals that are genetically more or less susceptible to infection. Agricultural studies have repeatedly shown that disease spread can be reduced in mixtures of susceptible and resistant plants as compared to a susceptible monoculture (e.g., Browning and Frey, 1969; Mundt and Browning, 1985; Wolfe, 1985). Among the many factors likely to be important are the smaller total number of susceptible plants in the mixture, the increased distance between susceptible plants (a plant density phenomenon), interference of resistant plants with disease spread between susceptible plants, and the existence of induced resistance (e.g., Burdon, 1978; Mundt and Browning, 1985; Wolfe, 1985).

Potential consequences. The question of how genetic variation persists in populations, that is, why a superior genotype does not exclude all others, is a major focus of attention for evolutionary biologists. Reduced disease spread in genetically variable plant populations may be one of the many factors that explain the maintenance of genetic diversity in natural populations in a way analogous to how density-dependent disease spread may explain the maintenance of species diversity in a plant community. In the case of intraspecific diversity, high densities of genotypes with the same susceptibility to a pathogen may have high levels of disease. A resistant genotype would have a higher probability of survival and reproduction near a susceptible genotype than would another susceptible individual. Thus the differential susceptibilities of different plant genotypes to disease, as well as the density-dependent process of disease spread, may maintain a wide array of genotypes in a population.

Diseases may also be an important force in the maintenance of genetic diversity in natural populations for other reasons. The ability of pathogens to track host phenotypes evolutionarily (Johnson, 1961; Barrett, 1981), partly due to both the discrepancy in generation times between plants and microbial organisms and the many ways that microorganisms exchange genetic information, can lead to rapid pathogen adaptation to hosts. A genetically uniform plant population would thus be extremely susceptible to exploitation by pathogenic organisms. The presence of disease could select for increased frequencies of any resistant plant genotype that appears due to mutation, sexual recombination, or immigration. The fitness of a plant genotype would, however, be frequency dependent, such that rarer plant genotypes would be more successful than common genotypes to which the pathogen population is already adapted. Theoretical models incorporating ideas of frequency-dependent selection of host resistance and pathogen virulence, and also potential fitness costs associated with these traits, have predicted the maintenance of genetically variable host and pathogen populations (Leonard and Czochor, 1980; Levin, 1982; May and Anderson, 1983).

Evidence. Evidence is needed to appraise two separate postulates: (1) that genetic variation for susceptibility to disease exists in natural populations, and (2) that such genetic variation is responsible for reduction in disease spread. Data on genetic variation for susceptibility to disease come from several sources using different approaches. Survey methods involve large-scale sampling of individuals from populations and assaying their disease resistance by artificially inoculating plants or by growing them in a nursery in association with inoculated plants that provide inoculum for natural infection. These methods are probably the most practical ways to approach analysis of large-scale spatial patterns in resistance, simply because so many individuals need to be tested to document patterns. All plants are also exposed to the pathogen, so disease escape is less likely to be confused with disease resistance. The disadvantage is that resistance screening tests may be done in sites geographically distinct from the original population, on individuals in monoculture, and with a small sample of pathogen genotypes; these features make it difficult to interpret the natural role of plant genotype on disease expression and epidemiology (Dinoor and Eshed, 1984). An alternative approach is the field transplant experiment, where individual plants are transplanted into a natural population and scored for subsequent infection in situ. For smaller-scale patterns in variation, the transplant experiment has the advantage that it actually tests the importance of plant genotype in disease expression with natural infection modes, natural pathogen genotypes, and in the context of natural variability in weather. Such studies may be limited, however, in the number of different genetic groups (i.e., plant clones, half-sib families, or full-sib families) that can be tested, since the variability inherent in field experiments necessitates large sample sizes per group.

Plant pathologists will be most familiar with disease resistance surveys done in populations of wild relatives of crop plants (e.g., Wahl, 1970; Zimmer and Rehder, 1976; Gerechter-Amitai and Loegering, 1977; Wahl et al., 1978, 1984; Segal et al., 1980). The purpose of many of these studies was to discover new resistance genes for use in breeding programs as well as to quantify the structure of the plant population in terms of resistance. The wild grasses of Israel have been particularly well studied, and both high levels of resistance and a diversity of types of its expression are found. A common approach for sampling the plant population is the transect method, as described by Wahl et al. (1978). For example, single heads of wild barley were collected at 1-m intervals in one study, regardless of the disease status of the plant. These seeds were sown in a nursery under heavy-mildew conditions. A great deal of variability was found in their susceptibility to disease, and in a total of 1700 plants assayed, over 60% showed some type of resistance reaction.

Using a survey method in a study of *Trifolium repens,* Burdon (1980) determined frequency distributions of resistance to a single isolate of each of two foliar fungal pathogens by inoculating clones (in the laboratory or greenhouse) derived from 50 individuals collected in a 1-ha grassland. A great deal of variability was found in the clones' reactions to both pathogens, although the frequency distribution of resistance differed for the two diseases. Burdon speculated that a frequency distribution skewed toward resistance may be the result of natural selection eliminating susceptible individuals; Dinoor and Eshed (1984) have pointed out that other factors (e.g., existence of tolerance to disease and environmental conditions) could also affect such distributions.

Two other studies by Burdon examined genetic variation at the population and species levels, by sampling several geographically diverse populations of closely related plant species. With four species of *Glycine* tested with a single isolate of the rust *Phakopsora pachyrhizi,* variation was found both among species and among and within populations of a single species in resistance of both a qualitative (absence of visible disease development) and a quantitative (varying degree of disease development) nature (Burdon and Marshall, 1981). In a study of populations of three wild oat species, variation in reaction to oat crown rust and oat stem rust was found using four isolates of each pathogen that were chosen for a wide range of pathogenicity (Burdon et al., 1983).

Others have instead focused on genetic variation present on smaller scales using transplant experiments. The rationale is to transplant plants of known genetic relationship back into their original population and examine disease levels from natural modes of infection on experimental plants. Work with the inflorescence disease of *Plantago lanceolata* caused by *Fusarium moniliforme* var. *subglutinans* was an example of this approach (Alexander et al., 1984). Plants from the field site were cloned and then maintained in an environmental chamber. Seed collected from these plants (half-sib families) were germinated and young seedlings were transplanted back into the field site in a randomized design. The transplants flowered the following year, and disease levels were observed to differ significantly among the plant families, suggesting that genetic factors accounted for some of the among-plant variability in disease levels observed naturally.

Parker (1985) performed a similar study with a native annual, *Amphicarpaea bracteata,* which can be infected by the fungus *Synchytrium decipiens.* His study not only explored whether or not genetic variation for disease susceptibility occurred in the host plant population but also the spatial character of fungal genetic variation. In his experiment, the families used were selfed progeny from seeds or seedlings collected in the field and grown to maturity in a common greenhouse. Six families, two derived from each of three sites, were transplanted into one of the original sites. The sites were spatially separated: two were 1 km apart and the third was 100 km from the other two sites. Disease levels were recorded on the experimental plants. Large differences in disease levels were found among the families, and in particular the site origin of the family proved to be important. Families transplanted into their native site had high levels of disease, while families from spatially separate sites were rarely affected. Since the parents of the plants used in the experiment were diseased in their original site, the apparent explanation is that the sites differed in the genetic composition of the fungal population. Fungal genotypes may have evolved the ability to infect local host genotypes, but were ineffective at attacking novel genotypes from other areas, even those found only 1 km away.

Parker's study illustrates the complexity inherent in the analysis of genetic variation in natural plant populations, since both host and pathogen population structure must be considered. In particular, his study suggests that selection may favor long-distance dispersal of plant propagules, which may lead to temporary escape from disease pressure.

The documentation of genetic variation for disease susceptibility in plant populations does not necessarily mean that such variation affects epidemiological parameters. Only a few studies of natural populations have specifically addressed the role of genetic diversity in disease epidemiology. Parker (unpublished data) studied a rust disease of mayapple *(Podophyllum peltatum),* a plant that forms genetically homogeneous local patches due to its clonal growth form. Despite high genetic uniformity, disease levels are low due to the presence of a morphological structure (scale leaves around the growing shoots) that protects the young plant tissue from contact with spores in the soil. Thus, with this plant, the lack of genetic variation in the plant population does not lead to high disease spread.

As discussed previously, the most extensive data on the types, frequencies, and spatial distributions of disease resistance in natural populations has come from the studies of wild progenitors of cultivated grains in Israel (e.g., Wahl, 1970; Wahl et al., 1978, 1984; Segal et al., 1980). Although there has been much interpretation of such data in terms of the role of genetic variation in protecting plant populations (e.g., Browning, 1974, 1980), little experimental work that tests this idea with these natural systems has been published. An exception is the work by Segal et al. (1982) with crown rust of oats, where plots in Israel were sown with either a monoculture of a susceptible oat cultivar, various Iowa multilines utilizing resistance genes from the Israeli wild oat relative *Avena sterilis,* a mixture of the recurrent parental lines of the multilines, or reconstituted populations of *A. sterilis.* The latter populations were created by first sampling natural populations of *A. sterilis* along transects, where seed of plants were collected at 1- to 2-m intervals regardless of the parent plants' disease reactions in nature. These seed were then planted in the experimental plots in the same spatial array as they were collected, thus producing populations that are similar to natural populations in both the degree and spatial patterning of genetic variation. Epidemics in the plots, as measured by aerial urediospore collections, were markedly lower on the multilines and reconstituted natural populations than on the susceptible cultivar or the recurrent hosts, suggesting that natural populations may be acting as natural multilines. In similar experimental plots, the numbers of races of *Puccinia coronata* f. sp. *avenae* were found to be higher and the proportion of a highly virulent race group were lower in the multilines and populations of *A. sterilis* than in the simpler mixture of recurrent hosts of *A. sativa* (Segal et al., 1982). Genetically variable host populations may therefore be affecting the genetic structure of the pathogen population as well as simply lowering the disease levels in the plot.

As intriguing as the results of this Israel study are, it is not a direct experimental test of the importance of genetic variation in reducing disease levels in populations of the wild *A. sterilis*. An important control was omitted, for there was no comparison of disease levels in populations of *A. sterilis* that are genetically variable (such as were reconstituted with the transect method) with populations of the same species that are genetically uniform (perhaps by planting tillers or seed obtained from one parent plant). The only study that I am aware of utilizing this approach to examine the epidemiological effects of host genetic variation in a nonagricultural setting was by Jarosz (1984). Individual plants of *Phlox* were identified with differential disease response to powdery mildew and experimental mixtures were created in the field with plants either 33%, 66%, or 100% susceptible. Inoculated plants were introduced in the center of the plots and the course of disease spread was followed on susceptible plants over the season. Contrary to prediction, no differences in disease epidemics (as measured by the area under disease progress curves) were observed among treatments. The author's explanation for his results focuses on the small size and low density of the experimental populations and the initial inoculum level, since initial dispersal of the fungus from the inoculated center was able to infect nearly all susceptible individuals in the plot. The presence or absence of resistant plants in the population thus had little effect on disease epidemiology. Further, the low plant density could also mean that autoinfections exceeded alloinfections, which can reduce the effectiveness of mixtures for disease control (Burdon and Chilvers, 1982).

If Jarosz's experiment had been done in larger plots or at a higher plant density, one might expect differences to appear among the treatments. Importantly, however, the author points out that the experimental design was consistent with the spatial scale of natural populations of *Phlox*. Unlike agricultural fields, many natural populations of plants, including *Phlox,* are found in small, separated populations. On such a spatial scale, intrapopulation genetic diversity may be a relatively unimportant factor in disease epidemiology, and thus our agricultural rationale for using multilines may not always apply to natural situations. This is of special interest since often multiline practices in agriculture are justified by being a natural method mimicking noncultivated situations.

Physical Environment

Prediction. *Spatial variation in the physical environment will affect disease spread.* It is of no surprise that physical factors of the environment, including light, moisture, and temperature levels, will have a large impact on disease incidence and disease spread. Such factors are likely to be much more variable, spatially, in natural as opposed to agricultural situations, and may account for variation in disease levels within and among plant populations.

Potential consequences. On an ecological level, a plant species' spatial distribution may be determined in part by disease levels if plants are much more susceptible to disease in certain habitats. Plant distributional patterns have been historically considered to be controlled by the ability of plants to tolerate various environmental factors, such as the light or soil regime, but the pathogens associated with these habitats may actually be the causal factors. Plants growing in environments highly conducive to disease may have high rates of infection and be poor competitors compared to other unaffected plant species, with the net result that distributional patterns are determined by disease.

A corollary to the above is that plants growing in environments that most favor disease will also have stronger selection for disease resistance than will plants grown in environments with low disease pressure. This may in turn lead to selection in pathogen populations for virulence, resulting in genetically diverse populations of both organisms.

Evidence. There are several studies that illustrate the role of physical factors in affecting disease levels. For many pathogens, including powdery mildew of *Phlox* (Jarosz, 1984) and damping-off diseases of several tropical trees (Augspurger, 1983, 1984; Augspurger and Kelly, 1984), both descriptive and experimental studies showed that disease levels were consistently higher in shady as opposed to sunny environments. Such patterns for tropical tree species are especially intriguing in that they suggest that pathogens may play an important role in forest regeneration. Recruitment of forest trees often occurs in gaps created in the forest canopy due to tree falls. The high light levels in the gap, as well as changes in soil conditions and plant–herbivore relations, are thought to determine colonization patterns (e.g., Hartshorn, 1978). Augspurger's work suggests that the reduced disease levels of the high-light environment may also be important. The exact cause for the lowered disease in sunny habitats is not known: it may be related both to the environmental requirements of the fungi and to the fact that seedlings in gaps grow faster and thus have a shorter time that they are vulnerable to damping-off diseases (Augspurger, 1983).

Testing for the existence of differential selection for plant resistance in different environments is a more difficult problem. Early workers recognized that disease resistance in wild grasses was most common in regions conducive to disease (Vavilov, 1938, cited in Wahl, 1970). Surveys of wild relatives of important grain crops have shown that spatial variation in plant resistance is often correlated with spatial variation in climate that could affect disease development (e.g., Wahl, 1970; Wahl et al., 1978). The work of Burdon et al. (1983) on wild oat populations in Australia and their rust pathogens also yielded similar results. Collections of seeds from 50 individuals were made from several populations in two regions of Australia: the northern region was in a warmer environment conducive to rust, while the southern region was much cooler and disease levels were generally low. Population collections were grown in the greenhouse and inoculated with a diverse collection of standard rust races. Higher levels of resistance were found for plant populations collected in the northern regions, suggesting that natural selection had acted such that only resistant genotypes were able to persist in regions with environments particularly conducive for disease.

Jarosz (1984) tested susceptibility to powdery mildew using two fungal isolates for 866 individuals of 10 taxa of *Phlox* collected from 112 populations in the central and south-central United States. For 65 of the populations, a habitat rating index was assigned based on tree cover and direction of exposure. This measure of degree of shadiness was significantly negatively correlated with host susceptibility at both the taxon and population level. Experimental studies by Jarosz (1984) also showed that artificial populations of *Phlox* grown in the shade had three to five times more diseased plants than similar populations grown in sunny areas. Jarosz concluded that populations that inhabited shaded environments have had higher disease pressure and have evolved higher levels of resistance than populations found in sunny habitats.

A contrasting view on the subject of relationships between disease resistance and the physical environment was provided by Dinoor and Eshed (1984). They mentioned a study on powdery mildew of wild cereals where resistance levels in the plant population were highest in a region not conducive to disease development. The authors stated that pathogens may behave more aggressively on susceptible plants in suboptimal environments, leading to selection for resistance. Data from this study, however, are not presented, so interpretation of these results is difficult.

Experimental studies are needed to clarify how selection for plant resistance depends on environmental conditions. Snaydon and Davies's (1972) research with the grass *Anthoxanthum odoratum* provides intriguing information on both the temporal and spatial scales that natural selection can act on in different environments. In a long-term study, grassland plots (of average size 17 m × 36 m) have received different fertilizer treatments since 1856 and different liming treatments since 1903. Apart from these treatments, plots have been managed identically. In their study, Snaydon and Davies collected tillers from the naturally occurring populations of *A. odoratum* in plots exposed to different nutrient treatments and planted them in a single plot. Despite being grown in a single environment, plants derived from the populations maintained under different experimental treatments differed for several morphological traits and for their levels of rust and mildew infection. Mildew resistance was greatest in populations from plots with high-nitrogen fertilizer treatments. Since high susceptibility to mildew often occurs when nitrogen fertilizer is used, the authors conclude that through the operation of natural selection, only resistant genotypes have been maintained in plots consistently exposed to high nitrogen. Rust resistance was highest in plots that naturally had tall vegetation. This may result from stronger selection for disease resistance in plots where humidity, which is necessary for infection, is kept high by the taller vegetation structure. Therefore, genetic variation for disease resistance present in the original population prior to the experimental treatments has been acted upon by natural selection within a time scale of 50 to 100 years and over distances as small as 30 m.

CONCLUSIONS AND RECOMMENDATIONS FOR FUTURE WORK

The reduced severity (or duration) of epidemics in natural populations has been attributed to such phenomena as "functional diversity," "ecosystem resistance" (Schmidt, 1978), or "protection of indigenousness" (Browning, 1974, 1980). These properties are thought to result from numerous characteristics of the host, pathogen, and environment, including the plant species diversity, diversity of types of disease resistance and tolerance within the plant population, diversity of types of virulence in the pathogen population, uneven plant age classes, and variable physical environments. Schmidt (1978) warns that diversity in these characteristics, in itself, does not guarantee low disease; instead, it must be diversity "of a kind that is functional against pathogens." It is clear from this chapter, however, that studies of plant–pathogen interactions in natural populations are still very incomplete; we especially lack experimental research that tests the importance of different factors on disease levels. Thus, although Schmidt presents interesting examples, currently the presence or absence of functional diversity can be identified only in an *a posteriori* sense by noting whether or not disease spread occurred given certain characteristics of the plant and pathogen populations. For example, few would have thought that large, diverse populations of American chestnut would be so completely devastated by the introduced pathogen *Endothia parasitica*.

Increasing the diversity of agricultural systems, for example by using mixtures of crops or cultivars, may prove to be very useful in disease control in agriculture; I do not wish to imply that we need to know the importance of a factor in natural populations before we apply it to crop populations. However, before we use the *rationale* that such methods be implemented because they provide a more natural means of managing disease, it makes sense to ensure that we have a detailed understanding of what factors affect disease epidemiology in natural populations. Increased understanding of disease dynamics in nature will require not only more studies of different plants and pathogens but scientific approaches specifically geared to integrate the fields of plant pathology, population ecology, and evolution. In particular, future work on disease in natural populations needs to overcome two major deficiencies: the lack of data on the effect of disease on plant fitness and, as mentioned, the lack of manipulative studies testing the importance of different biotic and abiotic factors on disease incidence and spread.

The effect of disease on the fitness of individual plants is crucial to understanding the effect of disease on the population dynamics of plants (Antonovics and Alexander, 1988). For example, on an ecological level, the patterns of survival and reproduction of individual plants with respect to disease levels may affect the population size and spatial distribution of a plant population. On an evolutionary level, the differential survival and reproduction of individual plants of different genotypes when exposed to disease could alter the genetic composition of the plant population. In many otherwise excellent studies of the types and spatial distributions of resistance in plant populations, however, no data or only anecdotal statements are given on the effect of disease on the fitness of individual plants. Although disease resistance is obviously related to plant fitness, the relationship can be very indirect, due to factors such as tolerance, when plants have high reproductive yields despite high disease levels.

As an illustration of why these data are so crucial, consider the classic paper of Browning (1974), which is often cited for demonstrating that genetic diversity protects natural populations from epidemics. Describing work on *A. sterilis,* Browning emphasized that resistance screening tests revealed that 70% of the population was susceptible to a common virulent race of *P. coronata* and thus that susceptible plants were not eliminated from the population. Several possible interpretations can be made from these data which cannot be distinguished without information from field studies of the effect of disease on fitness. For example, are the susceptible plants relatively free of disease in nature as a result of the protection given by the 30% of the population that are disease resistant? Or, are the susceptible plants rusted in nature, but the genotypes survive because of tolerance to disease (i.e., fitness is *not* affected by disease)? Yet another possibility is that in certain years and locations, susceptible plants are heavily rusted and selection for resistance exists. However, in other years or locations when disease pressures are less, the disease-free susceptible plants may be more fit than resistant genotypes, due to differences in growth or reproduction patterns, and thus both plant types are maintained.

Although we know more about resistance in the wild grasses of Israel than in any other system, we clearly still lack important information.

Studies of disease in natural plant-pathogen systems will also be enhanced by more experimental field studies. To test directly the effects of factors such as plant density, genetic variation, and physical factors on disease processes requires a manipulative approach. For example, despite the many papers that point to the role of genetic diversity of plants in maintaining low disease levels in nature, I could find only one study (Jarosz, 1984) that tested this hypothesis directly by examining disease spread in experimental field populations that were set up, in a relatively natural setting, to be either genetically uniform or variable. Field experiments, as opposed to greenhouse studies, also have the advantage of testing the importance of factors under more natural conditions in terms of inoculum levels and with the natural variability of weather. A critical test of any agricultural disease control strategy is the field experiment; similar methodology must also be carried over to the study of disease in natural populations.

Studies of disease in natural populations of plants must continue in the future. We can learn much about the basic ecological and evolutionary factors that mold antagonistic interactions between species from studies of parasitism. Theoreticians have invoked pathogens as being important in such diverse processes as the regulation of population size, maintenance of species diversity in communities, evolution and maintenance of sexual reproduction, and the maintenance of genetic variation in populations (e.g., Haldane, 1949; Chilvers and Brittain, 1972; Levin, 1975; Hamilton, 1980; May and Anderson, 1983; Rice, 1983). As obvious, however, few empirical and experimental data are available to test these predictions.

Besides our natural curiosity about interactions between pathogens and natural populations, there is information that can be gained to help with the practical problems of agricultural disease control. Plant pathologists have recognized and publicized the idea that natural plant populations may be a source of new ideas for disease control as well as their more traditional use as a source of resistance genes for breeding (Browning, 1974). Although the importance of genetic variation in disease spread in natural populations has been largely untested, the voluminous descriptive data suggesting that natural plant populations can be genetically diverse for resistance genes have played an important role in triggering agricultural interest in multilines and cultivar mixtures in the annual grain crops. Studies of disease in natural populations may be even more useful for comparison with agricultural situations that share many characteristics with natural populations, such as in many of the forage mixtures, tropical mixed cropping systems, and long-lived forest tree crops.

Regardless of the reason for studying disease in nature, our efforts will go farthest if communications are increased between the agricultural disciplines and the fields of ecology and evolutionary biology. Gould (1983) stated in a paper directed to applied entomologists and evolutionary biologists that "although researchers in both areas have occasionally quoted each other . . . there is a need for more active interaction between the two fields." The same is true for the diverse group of biologists interested in disease processes in nature. Much progress could be made if the artificial barriers separating plant pathologists and plant ecologists were removed to allow integrated in-depth research on the ecology and evolution of plants and their pathogens in natural populations. In turn, such studies could be used to provide a truly strong foundation for development of new methods of agricultural disease control.

ACKNOWLEDGMENTS

I thank J. Antonovics, C. Mundt, M. Parker, and A. P. Roelfs for their useful comments on earlier versions of this chapter. The National Science Foundation (grant BSR-8506734) provided support.

REFERENCES

ALEXANDER, H. M. 1984. Spatial patterns of disease induced by *Fusarium moniliforme* var. *subglutinans* in a population of *Plantago lanceolata.* Oecologia (Berlin) 62:141–143.

ALEXANDER, H. M., and BURDON, J. J. 1984. The effect of disease induced by *Albugo candida* (white rust) and *Peronospora parasitica* (downy mildew) on the survival and reproduction of *Capsella bursa-pastoris* (shepherd's purse). Oecologia (Berlin) 64:314–318.

ALEXANDER, H. M., ANTONOVICS, J., and RAUSHER, M. D. 1984. Relationship of phenotypic and genetic variation in *Plantago lanceolata* to disease caused by *Fusarium moniliforme* var. *subglutinans.* Oecologia (Berlin) 65:89–93.

ANTONOVICS, J., and ALEXANDER, H. M. 1988. The concept of fitness in plant–fungal pathogen systems. *In* Plant Disease Epidemiology, Vol. II (ed. K. J. Leonard and W. Fry), Macmillan Publishing Company, New York.

ANTONOVICS, J., and LEVIN, D. A. 1980. The ecological and genetic consequences of density-dependent regulation in plants. Annu. Rev. Ecol. Syst. 11:411–452.

ANONYMOUS, 1972. Genetic Vulnerability of Major Crops. National Academy of Sciences, Washington, D.C.

AUGSPURGER, C. K. 1983. Seed dispersal of the tropical tree, *Platypodium elegans,* and the escape of its seedlings from fungal pathogens. J. Ecol. 71:759–771.

AUGSPURGER, C. K. 1984. Seedling survival of tropical tree species: interactions of dispersal distance, light gaps, and pathogens. Ecology 65:1705–1712.

AUGSPURGER, C. K., and KELLY, C. K. 1984. Pathogen mortality of tropical tree seedlings: experimental studies of the effects of dispersal distance, seedling density, and light conditions. Oecologia (Berlin) 61:211–217.

BARRETT, J. A. 1981. The evolutionary consequences of monoculture. *In* Genetic Consequences of Man-Made Change (ed. J. A. Bishop and L. M. Cook), pp. 209–248. Academic Press, Inc. (London) Ltd., London.

BROWNING, J. A. 1974. Relevance of knowledge about natural ecosystems to development of pest management programs for agro-ecosystems. Proc. Am. Phytopathol. Soc. 1:191–199.

BROWNING, J. A. 1980. Genetic protective mechanisms of plant–pathogen populations: their coevolution and use in breeding for resistance. *In* Biology and Breeding for Resistance to Arthropods and Pathogens in Agricultural Plants (ed. M. K. Harris), pp. 52–75. Tex. Agric. Exp. Stn. Misc. Publ. 1451.

BROWNING, J. A., and FREY, K. J. 1969. Multiline cultivars as a means of disease control. Annu. Rev. Phytopathol. 7:355–382.

BURDON, J. J. 1978. Mechanisms of disease control in heterogeneous plant populations—an ecologist's view. *In* Plant Disease Epidemiology (ed. P. R. Scott and A. Bainbridge), pp. 193–200. Blackwell Scientific Publications Ltd., Oxford.

BURDON, J. J. 1980. Variation in disease-resistance within a population of *Trifolium repens.* J. Ecol. 68:737–744.

BURDON, J. J. 1982. The effect of fungal pathogens on plant communities. *In* The Plant Community as a Working Mechanism (ed. E. I. Newman), pp. 99–112. Blackwell Scientific Publications Ltd., Oxford.

BURDON, J. J., and CHILVERS, G. A. 1982. Host density as a factor in plant disease ecology. Annu. Rev. Phytopathol. 20:143–166.

BURDON, J. J., and MARSHALL, D. R. 1981. Inter- and intra-specific diversity in the disease-response of *Glycine* species to the leaf-rust fungus *Phakopsora pachyrhizi.* J. Ecol. 69:381–390.

BURDON, J. J., and SHATTOCK, R. C. 1980. Disease in plant communities. Appl. Biol. 5:145–219.

BURDON, J. J., OATES, J. D., and MARSHALL, D. R. 1983. Interactions between *Avena* and *Puccinia* species. I. The wild hosts *Avena barbata* Pott Ex Link, *A. fatua* L., *A. ludoviciana* Durieu. J. Appl. Ecol. 20:571–584.

CHILVERS, G. A., and BRITTAIN, E. G. 1972. Plant competition mediated by host-specific parasites—a simple model. Aust. J. Biol. Sci. 25:749–756.

CLAY, K. 1984. The effect of the fungus *Atkinsonella hypoxylon* (Clavicipitaceae) on the reproductive system and demography of the grass *Danthonia spicata.* New Phytol. 98:165–175.

CULLEN, J. M., and GROVES, R. H. 1977. The population biology of *Chondrilla juncea* L. in Australia. Proc. Ecol. Soc. Aust. 10:121–134.

DINOOR, A., and ESHED, N. 1984. The role and importance of pathogens in natural plant communities. Annu. Rev. Phytopathol. 22:443–466.

DINUS, R. J. 1974. Knowledge about natural ecosystems as a guide to disease control in managed forests. Proc. Am. Phytopathol. Soc. 1:184–190.

GERECHTER-AMITAI, Z. K., and LOEGERING, W. Q. 1977. Genes for low reaction to *Puccinia graminis tritici* in *Aegilops* and *Triticum.* Crop Sci. 17:830–832.

GOULD, F. 1983. Genetics of plant–herbivore systems: interactions between applied and basic study. *In* Variable Plants and Herbivores in Natural and Managed Systems (ed. R. F. Denno and M. S. McClure), pp. 599–653. Academic Press, Inc., New York.

HALDANE, J. B. S. 1949. Disease and evolution. Ric. Sci. Suppl. 19:68–76.

HAMILTON, W. D. 1980. Sex versus non-sex versus parasite. Oikos 35:282–290.

HARPER, J. L. 1977. Population Biology of Plants. Academic Press, Inc. (London) Ltd., London.

HARTSHORN, G. S. 1978. Tree falls and tropical forest dynamics. *In* Tropical Trees as Living Systems (ed. P. B. Tomlinson and M. H. Zimmermann), pp. 617–638. Cambridge University Press, Cambridge.

JANZEN, D. H. 1970. Herbivores and the number of tree species in tropical forests. Am. Nat. 104:501–528.

JAROSZ, A. M. 1984. Ecological and evolutionary dynamics of *Phlox-Erysiphe cicho racearum* interactions. Ph.D. dissertation, Purdue University.

JOHNSON, T. 1961. Man-guided evolution in plant rusts. Science 133:357–362.

LEONARD, K. J., and CZOCHOR, R. J. 1980. Theory of genetic interactions among populations of plants and their pathogens. Annu. Rev. Phytopathol. 18:237–258.

LEVIN, D. A. 1975. Pest pressure and recombination systems in plants. Am. Nat. 109:437–451.

LEVIN, S. A. 1982. Some approaches to the modeling of coevolutionary interactions. *In* Coevolution (ed. M. H. Nitecki), pp. 21–65. The University of Chicago Press, Chicago.

MAY, R. M. and ANDERSON, R. M. 1983. Parasite–host coevolution. *In* Coevolution (ed. D. J. Futuyma and M. Slatkin), pp. 186–206, Sinauer Associates, Inc., Sunderland, Mass.

MUNDT, C. C., and BROWNING, J. A. 1985. Genetic diversity and cereal rust management. *In* The Cereal Rusts, Vol. II (ed. A. P. Roelfs and W. R. Bushnell), pp. 527–560. Academic Press, Inc., Orlando, Fla.

PARKER, M. A. 1985. Local population differentiation for compatibility in an annual legume and its host-specific fungal pathogen. Evolution 39:713–723.

PARKER, M. A. 1986. Individual variation in pathogen attack and differential reproductive success in the annual legume, *Amphicarpaea bracteata*. Oecologia (Berlin) 69:253–259.

RICE, W. R. 1983. Parent–offspring pathogen transmission: a selective agent promoting sexual reproduction. Am. Nat. 121:187–203.

SCHMIDT, R. A. 1978. Diseases in forest ecosystems: the importance of functional diversity. *In* Plant Disease: An Advanced Treatise, Vol. II (ed. J. G. Horsfall and E. B. Cowling), pp. 287–315. Academic Press, Inc., New York.

SEGAL, A., MANISTERSKI, J., FISCHBECK, G., and WAHL, I. 1980. How plants defend themselves in natural ecosystems. *In* Plant Disease: An Advanced Treatise, Vol. V (ed. J. G. Horsfall and E. B. Cowling), pp. 75–102. Academic Press, Inc., New York.

SEGAL, A., MANISTERSKI, J., BROWNING, J. A., FISCHBECK, G., and WAHL, I. 1982. Balance in indigenous plant populations. *In* Resistance to Diseases and Pests in Forest Trees (ed. H. M. Heybroek, B. R. Stephan, and K. von Weissenbery), pp. 361–370. Pudoc, Wageningen, The Netherlands.

SNAYDON, R. W., and DAVIES, M. S. 1972. Rapid morphological differentiation in a mosaic environment. II. Morphological variation in *Anthoxanthum odoratum*. Evolution 26:390–405.

ULLSTRUP, A. J. 1972. The impacts of the southern corn leaf blight epidemics of 1970–1971. Annu. Rev. Phytopathol. 10:37–50.

WAHL, I. 1970. Prevalence and geographic distribution of resistance to crown rust in *Avena sterilis*. Phytopathology 60:746–749.

WAHL, I., ESHED, N., SEGAL, A., and SOBEL, Z. 1978. Significance of wild relatives of small grains and other wild grasses in cereal powdery mildews. *In* The Powdery Mildews (ed. D. M. Spencer), pp. 83–100. Academic Press, Inc. (London) Ltd., London.

WAHL, I., ANIKSTER, Y., MANISTERSKI, J., and SEGAL, A. 1984. Evolution at the center of origin. *In* The Cereal Rusts, Vol. I (ed. W. R. Bushnell and A. P. Roelfs), pp. 38–77. Academic Press, Inc., Orlando, Fla.

WOLFE, M. S. 1985. The current status and prospects of multiline cultivars and variety mixtures for disease resistance. Annu. Rev. Phytopathol. 23:251–273.

ZIMMER, D. E., and REHDER, D. 1976. Rust resistance of wild *Helianthus* species of the north central United States. Phytopathology 66:208–211.

8

Genetic Divergence in Spatially Separated Pathogen Populations

J. V. Groth and H. M. Alexander*

INTRODUCTION

Like all organisms, plant pathogens exist in populations that are influenced greatly by spatial aspects of their environment. For population-ecological questions in plant pathology, it is crucial to understand spatial heterogeneity of the environment and spatial separation of subpopulations, for these factors will affect both the evolutionary processes that select for particular genetic traits and the probability of interbreeding of genetically distinct members of the population.

The terms *genetic diversity* and *genetic divergence* are often used when describing pathogen populations in the context of spatial factors in the environment. The two terms will be used in the following way in this chapter:

Genetic Diversity. Genetic variation among individuals within a species. Genetic diversity is implied to underlie phenotypic diversity if the environment is ruled out as the cause of phenotypic diversity, but all genetic diversity does not necessarily result in phenotypic diversity.

Genetic Divergence. Genetic diversity caused by spatial or temporal separation of members of a species; a special, less random state of genetic

Note: Department of Plant Pathology, University of Minnesota, St. Paul, MN 55108, US. Department of Biology, University of Louisville, Louisville, KY 40292, US. *Current address: Departments of Botany and Systematics & Ecology, University of Kansas, Lawrence, KS 66045, US.

diversity. Both chance events and selective processes can lead to genetically divergent populations. In reality, it is often difficult to determine whether or not divergence, in the sense given here, has occurred in a population.

The foregoing definitions are probably restrictive in that they deal with the ideal concept of species as a group of organisms that exchange genes freely. Diversity and divergence can occur at other levels of taxonomy. At levels above species, these phenomena are part of speciation. Hierarchies of diversity also exist below the species level. In some of the cereal rusts and mildews, the special forms (based on ability to attack a host genus or higher taxon) versus the races within those special forms (based on the ability to attack a cultivar within the susceptible host species) provides an example which, although a possible artifact of modern agriculture, gives a clear hierarchical pattern of diversity which can also be considered to be an example of divergence.

It must also be borne in mind that the concept of species as a group of individuals that share the same gene pool is not well established for plant pathogenic microorganisms. Data are lacking. In fact, it may be that the discreteness implied in this concept of species is not as clear for microorganisms as it is for some or most of the higher organisms that inspired the concept of species. Thus divergence may be the rule for plant pathogens, and patterns of divergence and diversity may be complex in both space and time.

In this chapter we explore the concepts of genetic diversity and divergence in two ways. We first address the question of sampling pathogen populations, the first step in undertaking any population level study. Second, we focus on a specific pathogen, *Uromyces appendiculatus,* the cause of bean rust, and describe studies that illustrate the complex nature of research on genetic divergence of populations, especially as affected by spatial considerations.

SAMPLING PATHOGEN POPULATIONS

Before one can study the phenomenon of genetic divergence, one must address the basic question of how to sample from a pathogen population. Sampling considerations arise from more than just methodological problems. Different approaches to sampling can lead to dramatically different results and interpretations of the population ecology and genetics of the organism. As an illustration, we discuss sampling procedures used in race surveys of the airborne cereal rusts and identify and discuss some of the major biases that can occur.

Regional Sampling

It has been usual for samples of rust fungi taken from distant parts of large regions to be bulked together in collections. Such an approach is not inherently incorrect. In some parts of the world and for some pathogens, long-distance movement of pathogen propagules is the rule, so that it is partly justified to think of the pathogen in the entire epidemiological area as a single population. Bulking of samples is also, of course, easier for the investigator than handling them separately, since it reduces the total number of samples that must be maintained and characterized. Problems can, however, arise when making genetic and evolutionary interpretations about data obtained from bulked collections. Wolfe and Knott (1982) have addressed the problem of regional sampling with respect to virulence associations and disassociations, which are particularly important to measure in order to devise effective deployment strategies for highly specific resistance genes. They correctly point out the erroneous conclusions that can result from either bulking collections or prematurely bulking data obtained from separate collections. If two samples have contrasting frequencies of virulence to two host cultivars, combining them in one analysis can result in incorrect expectations and interpretations about the frequencies of isolates virulent on both host cultivars.

Analogous problems arise if bulked samples are used to attempt to characterize a pathogen population genetically. A simple example can perhaps best illustrate this. Suppose that one was interested in the question of whether or not a population was in Hardy–Weinberg equilibrium. Two samples were made at distant points and bulked, and through selfing of a large number of isolates from this bulked sample, the genotype of each at a single locus determining virulence to a host line was determined. The following hypothetical results were obtained:

frequency of genotype

AA	Aa	aa
0.41	0.18	0.41

frequency of gene

A	a
0.5	0.5

where the gene frequency of A is obtained as $f(A) = f(AA) + \frac{1}{2}(f(Aa))$.

The obvious conclusion to be drawn is that the population is not in Hardy–Weinberg equilibrium, since the frequency of heterozygosity is only 0.18 rather than the expected 0.50 derived from 2(0.5 × 0.5), the central term of the expanded binomial $(p + q)^2$. The scarcity of heterozygotes might be interpreted as a case of nonrandom mating (inbreeding) or as a case of homozygote fitness advantage.

Now suppose that the same two samples had been kept separate, with the following results:

sample 1 frequency of genotype			sample 2 frequency of genotype		
AA	*Aa*	*aa*	*AA*	*Aa*	*aa*
0.81	0.18	0.01	0.01	0.18	0.81
frequency of gene			**frequency of gene**		
A	*a*		*A*	*a*	
0.9	0.1		0.1	0.9	

In this case the conclusion is that both local populations were in Hardy–Weinberg equilibrium. Not only did bulking of samples from divergent populations of a pathogen result in an erroneous conclusion about genetic composition, it also failed to detect the divergence that gave rise to the misinterpretation. The example illustrates the fact that a certain amount of judgment based on familiarity with the pathogen environment (including the host) is necessary when decisions are made about how and where to sample and when samples can be bulked. Prior knowledge of the scale of genetic heterogeneity is necessary and, ideally, circularity of reasoning is avoided by understanding the environmental heterogeneity (patchiness) that contributes to genetic heterogeneity.

Host Selection

Another major sampling problem that has been documented, as well as alluded to in Wolfe and Knott (1982), is the indiscriminant bulking of samples from differentially resistant (''trap'') host lines which select out a portion of the population. More representative samples are expected when taken from uniformly susceptible lines. Browder (1966) has documented the possible size of the bias introduced by trap lines, which tend to overrepresent those virulences that specifically match the resistances present, and through displacement, also underrepresent phenotypes with other virulences since the sum of all phenotype (race) frequencies equals 1.

It should be stressed that the objective of representative sampling is not the same as the primary objective of much of the sampling that has been done in the past. Representative sampling, where all isolates of obligate parasites are obtained from uniformly susceptible hosts, allows estimation of frequencies of a range of traits. This is the first step in characterizing diversity. The original objective of monitoring certain virulences, which can be done effectively using biased sampling on resistant cultivars (trap plots), remains an important part of pathogen surveys. Trap plots also can add to the information obtained in representative samples by indicating whether detectable levels of virulence exist in a certain location. For example, we have used trap plots successfully with bean rust to give a primary screen of new differential bean lines. Those lines for which virulence or avirulence is 100% in a location are not considered further when polymorphism is the genetic condition of interest. Trap lines are much more likely to detect rare virulence than are representative sampling methods, especially if more than one generation is allowed so that very rare infections can multiply differentially. This can be accomplished most easily by planting late enough to allow young, susceptible host tissue to be present during the logarithmic phase of pathogen increase. The use of mobile nurseries, in which plants are taken to the field at the desired stage of growth, can accomplish the same result. If rust levels are to be quantified, trap plots or mobile nurseries containing differential lines must include susceptible cultivars to permit an estimation of the total rust population. For several reasons, however, this is not likely to give a very accurate estimation of frequencies of virulence. Different rates of increase of trap-selected versus total rust population, small differences in receptivity of host lines (Groth and Urs, 1982), and differences in leaf area or plant architecture are some of the variables that must be accounted for.

Representative sampling methods must be included in pathogen surveys, more so than in the past. These methods must also be suited to the specific objectives of the survey, and sampling patterns must be appropriately scaled. If within-field heterogeneity is suspected, a single sample from a small part of the field is not adequate. Mobility of the particular pathogen, environmental heterogeneity, the size and genetic diversity of the initial inoculum, and existing data on pathogen diversity are some of the parameters that should be examined in determining the spatial scale of sampling. These and other considerations were discussed by Allard (1970), who linked sampling scale to variance of the trait(s) of interest, such that traits that vary clinally should be sampled at intervals that are likely to reveal statistically significant frequency differences. Discontinuous or mosaic patterns must be handled differently. It is wise in such cases to consider hierarchical patterns of sampling, whereby both intra- and interpatch sampling combine to give a reasonably complete representation of the mosaic. It should be emphasized that proper sampling will always require preliminary knowledge of or assumptions about the diversity of the organism; each pathogen is likely to be different in this respect, considering the multiplicity of factors that can influence diversity and divergence.

GENETIC DIVERSITY AND DIVERGENCE IN *UROMYCES APPENDICULATUS*

With a background in some of the general problems involved with sampling pathogen populations to study genetic diversity and divergence, we present some specific studies of the ecological genetics of *Uromyces appendiculatus,* the fungus that causes bean rust. After introducing this organism, we first present descriptive studies, which have been the traditional approach used by plant pathologists to quantify genetic variation. Then we address experimental studies, which although performed in somewhat artificial environments, have advantages for studying the dynamics of evolutionary processes, most particularly, selection.

Description of *Uromyces Appendiculatus*

The causal agent of bean rust [*Uromyces appendiculatus* (Pers.) Unger] is an autecious rust species that has been transported around the world wherever its hosts (mostly *Phaseolus vulgaris*) are grown (Anonymous, 1981). The center of origin of the pathosystem is Central America, where about 30 or so species of wild *Phaseolus* and related genera occur that serve as hosts for the fungus (Laundon and Waterston, 1965). Information about the existence of host specialization above the species level and relationships with similar "species" of *Uromyces,* such as *U. vignae* on cowpeas, is very incomplete. Where both host specialization and morphological criteria do not support splitting a pathogen into multiple species, a single species is the safest designation. This course has been followed for *U. appendiculatus* by several workers (Arthur, 1934; Laundon and Waterston, 1965; Cummins, 1978). This means, however, that the patterns of genetic divergence will include levels of subdivision that might otherwise be excluded because of a splitter's philosophy. Our view is that species separation cannot be used to argue against seeking knowledge of genetic relationships of the allegedly distinct populations because, apart from the circularity inherent in the argument, the concept of species is too imperfect. In this chapter we concentrate on that part of the species of *U. appendiculatus* that can commonly be found only on cultivated *P. vulgaris,* because isolates from other hosts are currently lacking in our collections. It is likely that patterns of diversity and divergence will prove to be entirely different when the pathogen called *U. appendiculatus* is considered on all its host species, wild and cultivated. Studies of host-parasite specialization at various levels have been done for different reasons in the past. With economically important rusts of cereals, the emphasis has always been on the subpopulations of the pathogen found on crop species, at the expense of a more comprehensive picture. Anikster (1984) reviews the complexity of host specialization for rust fungi on cereals.

The type and extent of genetic divergence in bean rust are known only in a preliminary way. Many races of the bean rust fungus have been identified in various parts of the world. Since the first reports of races by Harter et al. (1935), the exact number of races cannot be enumerated because different sets of differential bean lines have been used in various places and times. In our opinion, the number of races is immaterial because of the genetic plasticity of the organism. We have chosen not to designate new races with race names in our work (Groth and Shrum, 1977). Suffice it to say that at least some of these races became predominant in a region because of spatial divergence relating to host specialization, and probably adaptation to local environmental conditions. For example, races from northern parts of the United States that overwinter as telia are specialized to attack the predominantly dry pinto and navy bean types, whereas a race that was collected from green beans in Wisconsin (Hagedorn and Wade, 1974), which apparently cannot produce telia, is more specialized to green or wax bean types (Groth and Shrum, 1977). This pattern of divergence can be explained and has been documented for many pathogens. There are other kinds of genetic diversity which cannot be explained so readily, and an indeterminable amount of this can be considered as genetic divergence. The description of this diversity is the topic of the remainder of this section.

Diversity within Local Populations of Sexual Bean Rust

Previous reports suggest that more than one race is commonly to be found in small field collections of bean rust (Groth and Roelfs, 1982; Stavely, 1984). We investigated this possibility using a single bulk collection of uredospores made in a small area of a field of cultivar Pinto 111 near Carrington, North Dakota, in September 1976. Pinto 111 is uniformly susceptible to all northern sexual isolates of bean rust, and thus use of the cultivar minimizes sampling biases due to host selection. Ten single-uredial isolates were randomly obtained from several plants of Pinto 111 inoculated with a sample of uredospores from the original collection. These isolates were used to inoculate 11 differential lines of beans. Table 1 presents the reactions obtained on five differential bean lines that responded to infection by the isolates with two or more distinct disease reactions. The other six differential host lines gave uniform reactions to all 10 isolates. There were six different races among the 10 isolates. None of the commercial cultivars of beans in the region possess resistance genes that occur in these five differential lines. Since there is little or no apparent spatial separation between these races, this diversity is probably not a case of divergence, at least on the macroscale of plant to plant or field to field. On a microscale, it is possible, but unprecedented in studies of path-

TABLE 1 Reactions of 10 Single Uredial Isolates of Collection P14 of Bean Rust from a Few Plants of a Pinto 111 Field Near Carrington, North Dakota, on Five Differential Bean Lines

	Isolate[a]					
Bean Line	6	8	1, 4, 10	7	2, 3, 9	5
B1349	F	F	F	F	F	C
Golden gate wax	F	F	F	F	C	F
US#3	F	I	I	I	I	I
Montcalm	I	I	I	C	I	I
Valley	I	I	C	I	C	C

[a]F, hypersensitive fleck resistance; I, reduced pustule size, nonnecrotic; C, fully compatible.

ogens, that some of these races are adapted to different microenvironments, such as old versus young leaves or different parts of the plant canopy. Our collecting methods would have bulked races that were differentially adapted to such specific niches.

The example above is presented in part to point out that divergence and nonspatial diversity are not always easy to separate. If some of the races had been obtained in collections from an adjacent county, it would be tempting to implicate an environmental basis for the race differences. More representative sampling would, however, determine whether two collection sites of interest were populated either qualitatively or quantitatively by different races. We have done preliminary studies to describe local differences in race composition. Table 2 illustrates that collections made from fields of susceptible cultivars of beans only a few miles apart can be subtly different. The frequencies were obtained by inoculating differential lines with large numbers of bulk-collected uredospores and counting numbers of (usually necrotic) resistant and (nonnecrotic) susceptible lesions. Again, these differences cannot be accounted for on the basis of resistance in the host.

Sometimes diversity for virulence can be considered as spatial divergence, at least in part. In 1982 in Renville County, Minnesota, two different "resistant" cultivars of beans were introduced and became moderately to heavily rusted late in the season. In other words, the rust population was able to adapt to both resistances simultaneously. The number of resistance genes in each cultivar is unknown, but the occurrence of rust isolates that can attack one but not the other of them indicates that the two cultivars do not possess the same gene or genes. Fleetwood navy, bred in Ontario, is one of the cultivars and comprised about 55% of the dry bean acreage in the region in 1982. The other cultivar is Olathe pinto, which was bred in Colorado and described as being resistant to prevalent rust races there. Table 3 presents data on frequency of virulence to the two cultivars in four collections of rust made in

TABLE 2 Percentages of Virulence in Two Field Collections of Bean Rust Made from Susceptible Pinto Bean Cultivars

	Population[a]	
Bean Line	P21	P22
	%	%
US#3	64	67
Early Gallatin	66	40
814	100	96

[a]P21 was from Pinto 111 in a field a few kilometers from the Pinto 114 field where P22 was obtained. Both pinto cultivars are susceptible to all local isolates of bean rust.

TABLE 3 Percentages of Virulence to Bean Cultivars Olathe and Fleetwood in Six Collections of Bean Rust Made in Minnesota

Collection	Source Cultivar	Virulence on Bean Cultivar (%)	
		Fleetwood	Olathe
1982			
P21	Pinto 111	7.0	12.8
P22	Pinto 114	9.0	98.1
01	Olathe	37.6	100
F1	Fleetwood	51.3[a]	100
1976			
P7	Pinto 114	0	0
P8[b]	Pinto 111	7.2	6.3

[a]Resistant reaction a small pustule.

[b]From a separate area 180 km to the north.

1982. As might be expected, bulk collections in fields of resistant cultivars contained higher frequencies of virulence to the "home" cultivar, being 100% for Olathe and 51% for Fleetwood. The percentage of virulence to Fleetwood was not 100%, presumably because the nature of the resistance permitted some reproduction by the "avirulent" portion of the population. As a differential cultivar, Olathe detected large differences between collections P21 and P22. Whether Olathe fields were more proximal to P22 than P21 is not known. The two 1976 collections of bean rust were included to observe whether detectable virulence occurred in Renville County or an adjacent area that was sampled prior to introduction of the two resistant cultivars. Collection P7 from Renville County had no measurable virulence, but P8 from the north had virulence to both cultivars. This suggests that at some time before 1982, virulence was introduced into Renville County from outside. Once again, however, the virulence to both cultivars was "unnecessary," in that it was present in the state prior to introduction of the resistance.

Indeed, small trap plots of differential host lines during 1983–1986 in the region where P8 was obtained have revealed that virulence to exotic differentials is the rule. Of 38 differential lines planted one or more years, virulence was detectable on 30. These differential lines were selected because inoculation with many isolates of the fungus suggested that they contained resistance genes that were different from one another. The lines also have no record of use in the area either directly or as sources of resistance in locally important cultivars. Many of the lines are from other countries. For multiple-year plantings, rust levels did not appear to change. For example, the cultivar Aurora was planted all four years and trace amounts of rust were found each year. These year-to-year comparisons are only qualitative, however, since absolute rust levels varied.

Diversity of Sexual versus Asexual Populations of Bean Rust

Populations of pathogens sampled in different areas may have different genetic compositions for a variety of reasons, including the selective actions of abiotic and biotic environments and the role of chance events affecting propagule dispersal. With *U. appendiculatus*, an additional factor is that the life history of bean rust is not necessarily the same in all locations. In particular, the sexual stage is not always present. *U. appendiculatus* is essentially obligately sexual annually in northern regions, where epidemiological patterns of rust spread indicate localized foci, presumably from aecia. It is not likely that uredia or urediospores can survive the northern winters. Teliospores, which do survive the winters, are produced in great abundance late in the growing season. Pycnia or aecia have been found several times in the field (Zaumeyer and Thomas, 1957; Jones, 1966; personal observations by first author), and we have induced their production by covering seedlings with dead bean plants containing telia. It is possible to germinate teliospores in large numbers in the laboratory and obtain pycnia in the greenhouse for genetic studies. Many of the rust fungus isolates in our collections, however, will not produce teliospores. Isolates lacking telia have been collected in northern regions, but their frequency is probably lower than in the tropics. Other isolates that we have worked with produce telia infrequently. In both central Mexico and Puerto Rico we have observed that isolates of rust from the same small collections can vary from those that fail to produce teliospores through to those that produce them readily in aging uredia. The different types can be intimately associated on bean leaves in tropical fields, as we have observed in both countries. How such a polymorphism is maintained is a mystery.

We have worked with a small sample of isolates from the midwestern United States representing the two extremes of the spectrum of telia production. Using both virulence and isozyme markers, some interesting patterns of diversity have begun to emerge. Isozyme data are taken from Lu (1984). These data are presented in Tables 4 and 5 for virulence markers and isozymes, respectively, with six isolates of bean rust. For the two kinds of markers, the patterns are similar for sexual versus asexual subpopulations. The sexual isolates are more homogenous than are the asexual isolates, and the two groups differ from one another considerably on average. It is among isolates of each of the two groups that a difference can be seen between the groups. The asexual isolates have lower coefficients of similarity than do the sexual isolates. This pattern is similar for both classes of markers even though the absolute degree of variation is not the same for the two kinds of markers used.

The general patterns support the view that members of a sexual population that undergo genetic recombination regularly are more homogeneous than are members of an asexual population that undergo genetic recombination

TABLE 4 Diallel Comparison of Six Isolates of Bean Rust for Proportion of Similarity in Virulence on 8 to 10 Differential Bean Lines

	Asexual[a]			Sexual[b]		
	W73-2	U2-1	S3-1	P10-1	S1-5	KW7-1
W73-2	—	0.50[c]	0.00	0.25	0.25	0.12
U2-1		—	0.10	0.20	0.50	0.10
S3-1			—	0.60	0.60	0.70
P10-1				—	0.60	0.90
S1-5					—	0.70
KW7-1						—

[a]Isolates that lack telia-producing ability.

[b]Isolates that produce teliospores readily.

[c]Proportion of markers examined that were not different in the two isolates.

TABLE 5 Diallel Comparison of Six Isolates of Bean Rust for Proportion of Similarity in Band Pattern of 14 Putative Isozyme Loci

	Asexual[a]			Sexual[b]		
	W73-2	U2-1	S3-1	P10-1	S1-5	KW7-1
W73-2	—	0.86[c]	0.37	0.56	0.56	0.59
U2-1		—	0.37	0.48	0.48	0.52
S3-1			—	0.78	0.78	0.78
P10-1				—	1.00	0.93
S1-5					—	0.93
KW7-1						—

[a]Isolates that lack telia-producing ability.

[b]Isolates that produce teliospores readily.

[c]Proportion of markers examined that were not different in the two isolates.

only infrequently. There is isolation and divergence between the two groups, but also within the asexual group.

Experimental Studies

The studies above provide a description of the genetic variation present in bean fields. The approach used, sampling populations and testing them on differential cultivars and with isozyme techniques, answers the question of *what* genes and genotypes are present. It is far more difficult to answer questions of *why* genetic differences are present or *how* divergence occurred and is maintained. Experimental studies, in which genetic change in pathogen populations is measured after exposure to a particular environment, can begin to answer these difficult, yet important questions.

In bean rust we have demonstrated that considerable genetic variations can exist on a small spatial scale in local populations. The next question, with important practical implications, is: What is the capacity of bean rust to adapt rapidly to the environment using existing, detectable variation? In a greenhouse environment, we have measured rates of change in frequency of several specific virulences (Alexander et al., 1985) of *U. appendiculatus* when maintained for five uredial generations on a partially resistant but, with respect to the virulence markers used, "neutral" host line (Fig. 1). The amount of change that was found in virulence frequency is striking. Of particular interest was the finding that changes in virulence to different cultivars did not all proceed in one direction. Instead, both increase in virulence (to US#3) and decrease in virulence (to Early Gallatin, Roma, and B1349) were found in the same population.

Previous studies where a large number of bean rust isolates were screened for virulence on the four differential bean lines indicate that at least three distinct resistance genes occur in the lines. Based on their reactions to many isolates of the fungus, Early Gallatin probably contains the same functioning single gene for resistance as is present in Roma, thus explaining the similarity in between-generation changes found for virulence to these two cultivars. Neither the magnitude nor direction of the changes of each virulence can be explained further, nor can the original polymorphism. However, the results clearly show that there was heterogeneity for overall fitness of individuals with different virulence traits in this population. Rapid shifts in virulence frequency are thus not only possible but probable when such populations are exposed to different environmental conditions. Sexual reproduction does occur annually in the population of bean rust represented by the sample in Fig. 1. The ubiquity and magnitude of change observed in this study during vegetative cycling suggests that Hardy–Weinberg equilibrium might not be expected in populations of uredospores that were obtained, as these were, at the end of the growing season; when several cycles of vegetative reproduction presumably separate the sexual reproduction event of early summer from the time of collection. This possibility is examined further below.

To examine how the sexual/asexual transition in the life cycle affects virulence dynamics, a study was made of change in virulence frequency in another sample of sexual bean rust from the same area and collected at the same time as the collection used in the selection study above. The difference in this study was that the collection was taken through the complete sexual cycle. Figure 2 shows the changes in frequency of virulence to four differential bean lines when a large field collection of uredospores was used to inoculate Pinto 111 and the uredia were allowed to convert to telia. Teliospores were germinated, and resultant pycnia on susceptible Pinto 111 were bulk-selfed.

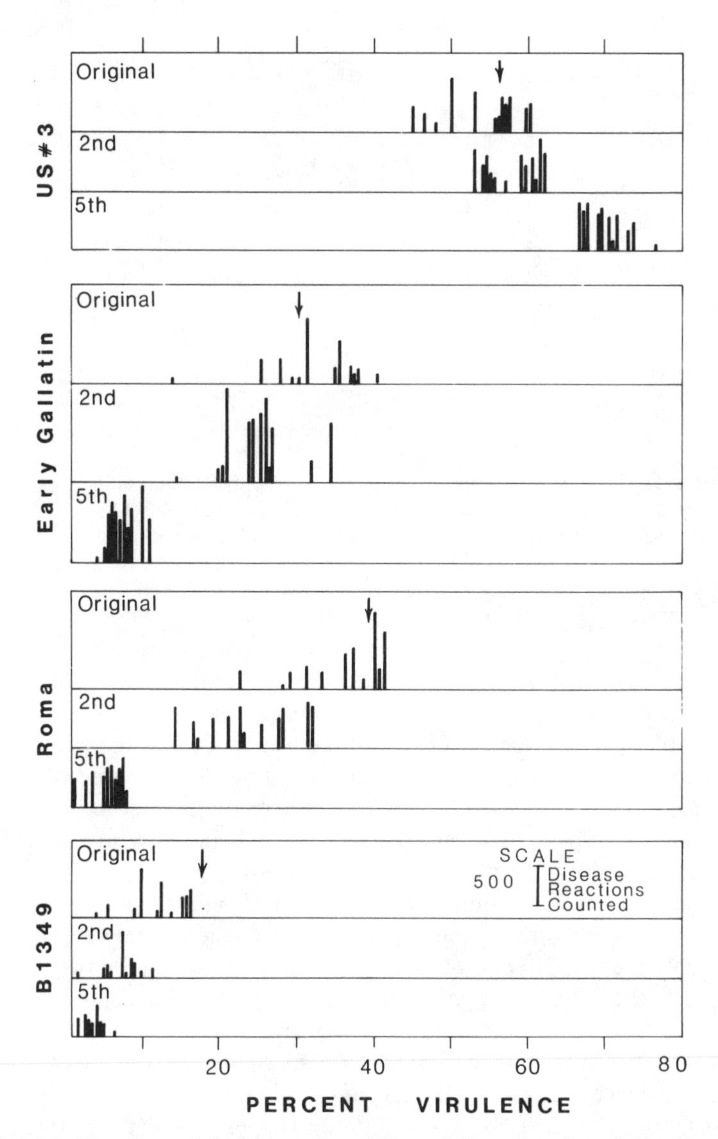

Figure 1 Frequency of virulence on four differential bean lines for a field-collected urediospore population of bean rust and for populations maintained on cv. Slimgreen (partially resistant) for two and five uredial generations. Each vertical line refers to a virulence frequency on one plant, with line length reflecting total counts. Arrows show initial frequency.

Change in frequency was generally greater in the vegetative part of the sexual cycle when aeciospores were used to inoculate plants (resulting in uredia) than in the actual sexual reproduction of teliospores to aeciospores. Changes in frequency of virulence were significant during the sexual reproduction only on

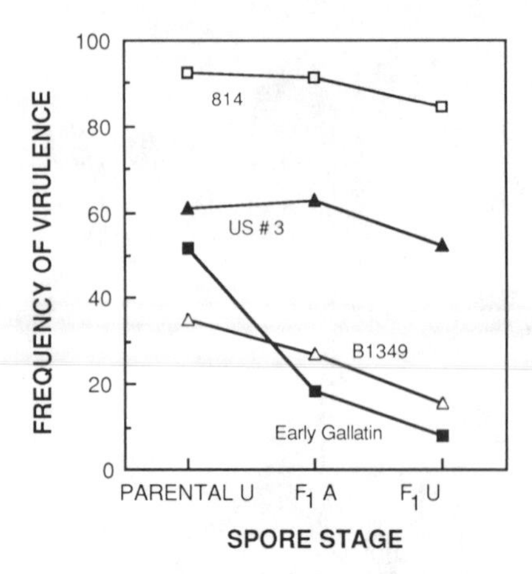

Figure 2 Changes in frequency of virulence in a field collection of bean rust urediospores (parental U) when increased as telia, teliospores germinated, and resultant pycnia formed on susceptible Pinto 111 bean leaves. Pycnia were mass fertilized to produce a large F_1 aecial (F_1A) population. Aeciospores and resultant urediospores (F_1U) from inoculation with aeciospores were used to inoculate the four differential bean lines.

Early Gallatin, whereas changes were significant during the vegetative transition of aeciospores to urediospores in all four cases. The lack in change in frequency of virulence during the sexual stage suggests that the population was close to a state of Hardy–Weinberg equilibrium for the three differentials. This contrasts with the results of Fig. 1, where the magnitude of frequency change of phenotypes might be expected to result in departure from Hardy–Weinberg equilibrium during uredial generations.

Hardy–Weinberg equilibrium is maintained only when there is no selection or migration perturbing gene and genotype frequencies. For a single locus, equilibrium is reached in one generation of random mating (Hartl, 1980). Population size (ca. 10^4 to 10^5 pycnia) of our pycnial stage and the bulk method of mating pycnia approximates a large random mating event about as well as is practically possible. The lack of significant change in frequency of the "allele" for virulence [definitely established as a single allele only for the virulence to Early Gallatin (Christ and Groth, 1982)] suggests that (1) the frequency of heterozygosity is roughly similar to the Hardy–Weinberg expectation; and (2) selection is not operating differentially on the three genotypes of the "locus" for virulence. Change in frequency during the random mating event, such as was observed for virulence to Early Gallatin, could be caused by either or both of conditions 1 and 2 not being met. These generalizations should apply to virulence due to more than one locus provided that the loci are not linked. Epistasis, which is to be expected with genes for specific virulence, should also not interfere with the establishment or maintenance of Hardy–Weinberg equilibrium. This extends the logic to include many but not all of the cases of more complex modes of inheritance of virulence, such as that found by Christ and Groth (1982) involving bean line US#3, or that found by Kolmer (1982), where apparently two loci in bean rust interacted to condition virulence to bean line 814. In view of this, the most important information to be sought about multiple loci that interact to condition virulence is whether or not they are linked.

The changes observed in Fig. 2 during the aecial-to-uredial stage were all significant and all in the same direction—toward less virulence. This generally agrees with the results of uredial-to-uredial selection in Fig. 1 (except for virulence to US#3), but so many of the details of the two studies were different that comparison is difficult and interpretation perhaps premature. Rather than emphasize the direction of change, it is perhaps more useful to note the large magnitude of response to selection, although the selection pressure is not known or controlled. That there is great potential for rapid change in bean rust is clear from both studies of virulence change in bean rust.

CONCLUSIONS

There has been a clear trend in biology toward consideration of species of organisms from a population rather than a historical typological perspective. The transition is far from complete, and in fact it is unlikely that biologists will ever be able to consider species purely out of consideration of gene flow patterns. Such an approach will require detailed knowledge of the population genetics and ecology of each organism, and will probably result in a cumbersome system of nomenclature that will poorly serve those who work with the organisms. A compromise between these views will be needed, whereby the naming and discrete categorizing of organisms is based on a reasonable amount of knowledge of their population and ecological genetics. The artificiality and imperfections of the categories will then be appreciated, yet accepted. Those who use the names will know the limits of use of the groupings.

Once the issue of what constitutes a valid species is resolved, the question of intraspecific diversity must be addressed. Experience with better-known pathogens such as the rust fungi would suggest that rather than assuming by default that a plant pathogenic species of microorganism is invariate and has not undergone spatial divergence, the *a priori* assumption should be that it is diverse. Establishment of the nature and practical implications, if any, of diversity and divergence will then be the goal of the researcher.

As defined above, knowledge of spatial divergence of plant pathogens is more valuable than simple knowledge of diversity because in uncovering divergence one is likely also to uncover a certain amount of understanding of the underlying forces that caused the divergence. So often with diversity, the existence of differences is measured and becomes an indisputable fact, but the basis for this diversity is little more than a subject for speculation.

Once knowledge of diversity or divergence in a pathogen is obtained, a number of pragmatic changes by plant pathologists and plant breeders should follow. For plant pathologists, collection and maintenance of representative samples of the pathogen are complicated. A single isolate is no longer adequate, and the issue of representative sampling is introduced. Adequate screening nurseries for resistance breeding must also contain representative samples of the pathogen in the region in which the crop is to be grown. Beyond this aspect of breeding, plant breeders must use the knowledge and nature of pathogen diversity to decide what kinds of resistance to use and possibly how it is to be deployed. A diverse pathogen may be able to adapt to resistance quickly enough to render its incorporation cost ineffective. We have yet to reach that stage of knowledge of diversity of any pathogen that permits confident prediction of the durability of resistance.

Finally, the patterns of diversity may reveal details of the life cycle of the pathogen that are important in choosing between alternative or combinations of control strategies. Some examples would include information on direction and timing of seasonal or long-term pathogen migration as revealed by phenotype distribution in space and/or time, the importance of overwintering or oversummering of segments of a pathogen population, and the frequency of occurrence and implications for diversity of sexual or parasexual cycles. So although information about diversity and divergence is useful in itself, further interpretation may provide the insight that allows principles to be formulated about present and future behavior of pathogen populations.

REFERENCES

ALEXANDER, H. M., GROTH, J. V., and ROELFS, A. P., 1985. Virulence changes in *Uromyces appendiculatus* after five asexual generations on a partially resistant cultivar of *Phaseolus vulgaris*. Phytopathology 75:449–453.

ALLARD, R. W. 1970. Population structure and sampling methods. *In* Genetic Resources in Plants (ed. O. H. Frankel and E. Bennett), pp. 97–107. F.A. Davis Company, Philadelphia.

ANIKSTER, Y. 1984. The formae speciales. *In* The Cereal Rusts, Vol. 1: Origins Specificity, Structure and Physiology (ed. W. R. Bushnell and A. P. Roelfs), pp. 115–130. Academic Press, Inc., Orlando, Fla.

ANONYMOUS. 1981. Distribution maps of plant diseases. Commonwealth Mycological Institute, No. 290, *Uromyces appendiculatus* (Pers.) Unger.

ARTHUR, J. C. 1934. Manual of the rusts in the United States and Canada. Purdue Research Foundation, West Lafayette, Ind.

BROWDER, L. E. 1966. A rapid method of assaying pathogenic potential of populations of *Puccinia graminis tritici*. Plant Dis. Rep. 50:673–676.

CHRIST, B. J., and GROTH, J. V. 1982. Inheritance of virulence to three bean cultivars in three isolates of the bean rust pathogen. Phytopathology 72:767–770.

CUMMINGS, G. B. 1978. Rust fungi on Legumes and Composites in North America. The University of Arizona Press, Tucson, Ariz.

GROTH, J. V., and ROELFS, A. P. 1982. Genetic diversity for virulence in bean rust collections. Phytopathology (Abstr.) 72:982.

GROTH, J. V., and SHRUM, R. D. 1977. Virulence in Minnesota and Wisconsin bean rust collections. Plant Dis. Rep. 61:756–760.

GROTH, J. V., and URS, N. V. R. R. 1982. Differences among bean cultivars in receptivity to *Uromyces phaseoli* var. *typica*. Phytopathology 72:374–378.

HAGEDORN, D. J., and WADE, E. K. 1974. Bean rust and angular leaf spot in Wisconsin. Plant Dis. Rep. 58:330–332.

HARTER, L. L., ANDRUS, C. F., and ZAUMEYER, W. J. 1935. Studies on bean rust caused by *Uromyces phaseoli typica*. J. Agric. Res. 50:737–759.

HARTL, D. L. 1980. Principles of Population Genetics. Sinauer Associates, Inc., Sunderland, Mass.

JONES, E. D. 1966. Aecial stage of bean rust found in New York State. Plant Dis. Rep. 44:809–814.

KOLMER, J. A. 1982. Genetic studies of host interaction with dicaryotic and haploid stages of *Uromyces phaseoli* var. *typica*. M.S. thesis, University of Minnesota.

LAUNDON, G. F., and WATERSTON, J. M. 1965. *Uromyces appendiculatus*. CMI Descr. Pathog. Fungi Bacteria 57.

LU, T. H. 1984. Electrophoresis and isozyme genetics of bean rust (*Uromyces appendiculatus*). M.S. thesis, University of Minnesota.

STAVELY, J. R. 1984. Pathogenic specialization in *Uromyces phaseoli* in the United States and rust resistance in beans. Plant Dis. 68:95–99.

WOLFE, M. S., and KNOTT, D. R. 1982. Populations of plant pathogens: some constraints on analysis of variation in pathogenicity. Plant Pathol. 31:79–90.

ZAUMEYER, W. J., and THOMAS, H. R. 1957. A monographic study of bean diseases and methods for their control. USDA Agric. Res. Serv. Tech. Bull. 868:34–42.

9

National and International Breeding Programs and Deployment of Plant Germplasm

New Solutions or New Problems?

D. Marshall

INTRODUCTION

When we consider germplasm deployment and its subsequent adoption by producers, we focus explicitly on the changing area of utilization of the germplasm over time. The diffusion of germplasm over an area is determined by a set of factors that include the genetics of plants, the genetics of pathogens, and their interactions with each other and the environment. Because of the unique relationship between genetic resistance to pathogens, the area over which resistant cultivars are grown, and the pathogens themselves, I shall use the term *spaciogenic* to better describe the spatial phenomena of germplasm diffusion. Three geographic scales are important in spaciogenic aspects of agricultural systems: (1) intrafield considerations, where genetically similar or dissimilar plant genotypes are spatially distributed in a field; (2) interfield considerations, which account for the spatial separation of fields in an area; and (3) interregional factors, which deal with particular kinds of germplasm that are limited to specific regions (Mundt and Browning, 1985)

Plant breeders and pathologists have long been involved with the devel-

Note: Texas Agricultural Experiment Station, Texas A&M University Research and Extension Center, Dallas, TX 75252, US.

opment of high-yielding, disease-resistant germplasm. But little attention has been given to the nature and consequences of how germplasm should best be deployed. Germplasm is often deployed indiscriminately and then allowed to find its niche. As Buddenhagen (1983) noticed, this may be because it is an exception to find plant pathologists participating directly in the release of disease-resistant germplasm. Germplasm is most often released from a breeding program along with a recommended area of production that is based primarily on breeders' plot field trials, where the germplasm had outyielded currently grown cultivars (Marshall et al., 1986). Even though these small-plot field trials are conducted over a number of years and locations, the actual amount of the germplasm tested is small compared to commercial production of the germplasm. Therefore, the germplasm is exposed to a relatively small number of potential pathogens, pathotypes, and environmental stresses. As a result, germplasm that seemingly has wide geographical adaptation may be quite limited to a specific area of adaptation.

The conventional breeding approach, especially in the cereal crops, has been to develop genetically uniform, pure-line germplasm. Pure-lines may be well adapted to a given set of biological and environmental factors, but they are unable to adjust when the set of factors change (Simmonds, 1962). It is the genetically static nature of pure-lines over space and through time that increases the probability of disease epidemics. Within most of the cereal-growing areas of the world are planted cultivars with similar genetic backgrounds, over large, contiguous areas (Harlan, 1972; NAS, 1972). In most instances, disease epidemics, ephemeral disease resistance, and the continuous replacement of cereal cultivars can be attributed to the extensive cultivation of pure-lines and the failure to direct their spatial diffusion over time (Johnson, 1961; Harlan, 1972). In this chapter we focus on wheat and, to a lesser extent, other cereals because of the pivotal role that breeding and germplasm deployment have played in the culture of these crops.

For the majority of time that man has been engaged in growing cereals, the crops have been spaciogenically diverse on all the aforementioned geographic scales (Jasny, 1944; Harlan, 1975). In North America, Australia, and Europe, it has only been in the past 80 to 100 years that spaciogenic uniformity was customary. Mechanization and large-scale specialization greatly decreased interfield and interregional diversity (Marshall, 1977), while the pure-line method of breeding cereals virtually eliminated intrafield diversity (Allard and Hansche, 1964; Harlan, 1972).

In some less developed countries of the world, a more abrupt shift was taken from the spaciogenic diversity of traditional agriculture employing landraces over to the production of pure-line cultivars grown under more intensive cultivation. The shift, in wheat, resulted from the timely development and rapid adoption of the semidwarf, photoperiod-insensitive germplasm developed in Mexico in the 1950s and early 1960s. The combined success of the semidwarf germplasm and the associated high-yield technology resulted in the

Green Revolution (Borlaug, 1981). This revolution made it possible to improve the living conditions of much of the world's population in the less developed countries. Without question, the benefits of the Green Revolution were, and continue to be, tremendous. The fact remains, however, that diseases are still a major component in the failure of cereals to yield their genetic potential. Furthermore, it is the cereal rust diseases (the diseases that pure-line cultivars were originally bred to control) that still cause epidemics throughout the world. As Borlaug (1981) aptly pointed out: "The current methods of relying on pure-line, disease resistant cultivars can be improved upon." The Green Revolution set the example. It must now be used as the starting point in the development of new, sustainable breeding and deployment systems. The question is: Can pathologists and breeders reestablish spaciogenic diversity, eliminate disease epidemics, provide durable disease resistance, and subsequently reduce the evanescent nature of cereal cultivars while maintaining and improving current production levels?

To answer this complex question, a reflection on past events is first needed in order to examine how and why wheat cultivation went from a state of spaciogenic diversity to spaciogenic uniformity and the effects that this has had on wheat diseases.

HISTORICAL PERSPECTIVE

The first human consumers and users of wheat were those neolithics who gathered the plants from natural, mixed populations. After some migration, the early farmers sowed and later harvested mixed cereals in tiny plots (Ucko and Dimbleby, 1969). This simple farming system probably began and slowly moved away from an area near the present-day borders of Iran, Iraq, and Turkey (Clark, 1969). The first cultivated small grains were wild einkorn *(Triticum boeticum)*, wild emmer *(T. dicoccoides)*, and wild barley *(Hordeum spontaneum)*. Interestingly, these species may originally have been grown for their vegetative biomass for use as animal fodder rather than for grain (Bohrer, (1972). From about 3000 to 1850 B.C., the size of individual fields is fairly unknown, but they were necessarily small because farmers could only work as much ground as their primitive tillage implements and animals permitted (Fussell, 1965). As a result, the spatial separation between fields was large. Within fields, spaciogenic diversity between plants was high and the crops probably resembled the diverse cereal populations now found in natural stands in parts of the Middle East (Vavilov, 1951; Wahl et al., 1985).

From the writings of Cato (Brehaut, 1933) we learn something about the size of agricultural fields in the Mediterranean region from 500 B.C. to A.D. 500. The commercial crops of the time were olives and grapes. The average vineyard was 100 *jugera* (25.3 ha) in size. However, wheat fields were much smaller, ranging from 2 to 7 *jugera* (0.5 to 1.7 ha), and were grown in family, subsistence farming operations. The wheat and other cereal crops grown during this period were the archetypal landraces. The landraces were genetically heterogeneous populations that had adapted over the years to a changing biological environment, local soil types, climatic fluctuations, and cultivation practices (Dorst, 1957). Landraces were very reliable inasmuch as they produced fairly consistent yields from year to year. Most historical accounts relate that the yields of landraces were rather low (Vavilov, 1951; Dorst, 1957). However, records show that in central Iraq in 2400 B.C., wheat yields averaged 72 bushels of grain per hectare (Jacobsen and Adams, 1958). This can be compared with the wheat yields in Texas in 1980 of 62 bushels per hectare! Most important, the landraces enabled subsistence farmers to survive and persist (Lozano and Schwartz, 1981). Both qualitative and quantitative forms of disease resistance existed in the landraces (Vavilov, 1951; Qualset, 1975; Damania, 1985; Negassa, 1987). The individual components of a landrace were adapted not only to the environment but to each other as well (Witcombe and Gilani, 1979; Damania et al., 1985). Therefore, within the prevailing conditions of a region, in a given year and field, there would be some plants that compensated for others that were at a disadvantage. In the next year or field, the stresses might be different, and other components of the landrace population would reproduce differentially. Thus the landraces possessed a high degree of intrafield, spaciogenic diversity that helped protect them from disease epidemics (Harlan, 1975). Additionally, the landraces were adapted to the cultural conditions of traditional agriculture, namely low soil fertility and low plant populations. As a result, "crowd diseases" such as the rusts were inconsequential.

The landraces continued to evolve and adapt through the centuries to become local varieties or cultivars. They were still genetically heterogeneous and generally variable in appearance, but each was identifiable and locally named (Hunter and Leake, 1933). In the eighteenth century mechanized row cropping and seed drilling helped to stimulate wheat expansion in western Europe and North America. Stem rust was epidemic on wheat in Italy in 1766, particularly in the district of Tuscany (Fontana, 1767). Fogs during the cool nights followed by hot, sunny days provided optimal environmental conditions for infection by *Puccinia graminis tritici*. Tozzetti (1767) wrote that during the Tuscany stem rust epidemic, the wheat in the wheat–rye and wheat–vetch mixtures had less severe stem rust than did the wheat in pure stands. Apparently, given a highly conducive environment, rust epidemics could occur on the local cultivars of the mid-eighteenth century.

The epoch from about 1850 to 1910 had a great impact on wheat production, especially in the United States. Large areas of land in the Great Plains were being sown to wheat because land was cheap and available and farm mechanization was ever improving (Fussell, 1965). Interfield and interregional diversity began a rapid decline. Wheat was becoming spatially contiguous from Texas to the Canadian provinces of Alberta and Saskatchewan. In 1877, the foundation of the hard red winter wheat industry was established in the Great Plains with the introduction of the Turkey landrace from southern Russia by Mennonite immigrants (Salmon et al., 1953). The Turkey landrace was rather uniform in physical appearance, but was variable for many physiological traits, including disease resistance. In 1887, the U.S. Congress passed the Hatch Act, which provided the framework for public support of agricultural research and created the agricultural experiment stations. Much of the early work of wheat researchers in the state experiment stations in the Great Plains involved testing different lots of Turkey and selections from it and other landraces (Salmon et al., 1953).

From the late nineteenth century to 1920, most of the improvement in wheat was by means of introductions and subsequent selection. In the United States, the use of selection to improve wheat dates back at least 100 years before the Hatch Act, to 1787, when a hard winter wheat was selected from an unknown landrace in Virginia (Destler, 1968). Some researchers in the 1890s began to cross lines that they had selected from the landraces. The progeny of the crosses were allowed to self-pollinate over several generations and individual head selections were made. The first extensively grown pure-line cultivar of hybrid origin in North America was the hard red spring wheat Marquis, developed by C. E. Saunders of Canada and released in 1912. Marquis resulted from a cross between a selection from Hard Red Calcutta, an Indian landrace, and Red Fife, a selected landrace of Gaelic origin (Salmon et al., 1953). The farmers' acceptance of Marquis was slow at first, but its area of production increased rapidly, until by 1929, it occupied 90% of the northern Great Plains spring wheat area.

Generally, the adoption by producers of new germplasm, or any new technology or innovation, follows a logistically-shaped curve (Luce and Raiffa, 1958). The pattern of adoption can be defined by the time when the germplasm is introduced and the rate of adoption (Byerlee and de Polanco, 1982). Once germplasm is released, its acceptance as a cultivar is often slow at first. But if the cultivar performs well during the first years following release, its acceptance by producers and its acreage undergo a phase of rapid increase. The acreage planted to a cultivar levels off and often decreases following shifts in the pathogen populations and/or the release of new and improved cultivars.

Gould (1965) described the acceptance of breeder-recommended cultivars by farmers in Kilimanjaro, Tanzania, as a process of learning with partial reinforcement. This means that even though the acceptance of new cultivars decreased the probability of sustaining damage from rust, such acceptance did not ensure a good crop because of other production risks. As shown in Fig. 1, prior to 1943, less than 5% of the Kilimanjaro wheat-growing area was planted to recommended pure-line cultivars, because local cultivars and landraces predominated. When more virulent forms of *Puccinia graminis tritici* arose, the pure-line cultivars were found to have major gene resistance that

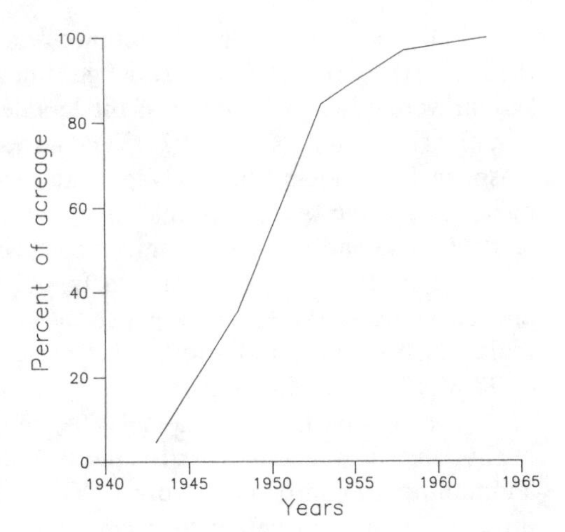

Figure 1 Percent of wheat acreage planted to breeder-recommended wheat cultivars in Kilimanjaro, Tanganyika, from 1943 to 1963 (after Gould, 1965).

protected their yield; therefore, farmers recognized those cultivars as superior and grew them in subsequent years. However, as the area of the pure-lines increased, selection pressure was placed on the *P. graminis tritici* population, which resulted in an increase of new races able to overcome the resistance in the pure-lines. As a result, the wheat farmers were caught in a boom-and-bust cycle. By 1963, essentially all the acreage was planted to breeder-recommended cultivars because other alternatives were unavailable.

Most of the wheat breeding programs in the world pursued similar courses of action from the 1940s through the 1980s. Pure-line breeding and, when needed, incorporation of major-effect genes for rust resistance were the norm. The recommended cultivars were planted on adjacent, expansive acreages. Many new cultivars were developed, released, and subsequently lost, mainly because of adaptation in the rust populations.

As wheat pure-lines and their associated seed industry intensified, some researchers were questioning the practicality of dependence on intrafield spaciogenic uniformity in self-pollinating cereals. Harlan and Martini (1929, 1938) pioneered early work on increasing genetic diversity in cereals by means of cultivar mixtures. Jensen (1952) promoted intrafield diversity as a means of obtaining stability in oat cultivation. Borlaug (1953) and Borlaug and Gibler (1953) reported on the development of spaciogenetically diverse multilines as a means of controlling stem rust epidemics in wheat. Borlaug's work on multilines at CIMMYT (International Maize and Wheat Improvement Center) was concomitant with research on semidwarf, photoperiod-insensitive germplasm. The multilines were based on standard-height germplasm that did not have the yield potential equivalent to that of the semidwarfs under intensive cultivation, as yields of 90% or better than standard-height cultivars were realized with the semidwarfs (Borlaug, 1981).

Following widespread acceptance and phenomenal success in Mexico in the 1950s, the CIMMYT semidwarf germplasm and its intensified management system were adopted in many of the less developed countries in Asia and Africa (Saari and Wilcoxson, 1974). The adoption curve of the semidwarf germplasm in Bangladesh, India, Nepal, and Pakistan (Fig. 2) indicates that the most rapid rate of adoption was between 1966 and 1971. This can be compared to the rate of adoption of semidwarf germplasm in Texas during the same period (Fig. 2). The Texas semidwarfs, developed mainly by K. B. Porter and his colleagues (Atkins, 1980), experienced a rapid acceptance just after their initial introduction, and another large increase in 1980 following the release of TAM 105 wheat in 1979.

Throughout the 1970s and 1980s, cereal breeding by private companies in the United States increased rapidly, both in the number of lines released and number of companies involved. The Plant Variety Protection Act of 1970 provided greater incentive to private companies to breed and release cereal cultivars. It is to the advantage of the commercial breeders to develop cultivars that will later need replacement, since it is much easier to promote a new cultivar with some improvements than to extoll the virtues of an already cultivated line. Hybrid small grains, whether produced cytoplasmically or by chemical hybridizing agents, advance the purchasing of new seed to an annual occurrence. The benefits of hybrids to commercial concerns are obvious. Spaciogenic diversity has not been a high priority among private concerns, nor, for that matter, public institutions.

It is evident that pure-line development and release has received most of the attention of small grain breeders from the very initiation of breeding programs. Although the present system has several positive attributes, some difficult problems, particularly related to genetic vulnerability to disease, have been created (NAS, 1972).

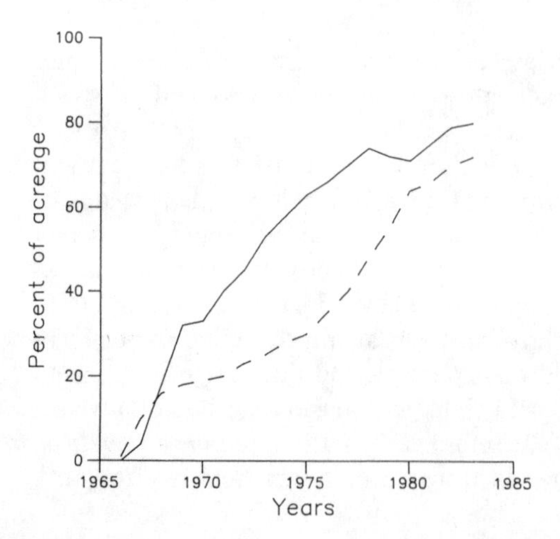

Figure 2 Percent of wheat acreage planted to semidwarf wheat cultivars from 1966 to 1983. The solid line represents semidwarf wheats in Bangladesh, India, Nepal, and Pakistan (after Dalrymple, 1986). The dashed line represents semidwarf wheats in Texas.

NEW PROBLEMS

One of the most underrated phenomena in agriculture is the introduction or intensification of new and unsuspected problems associated with changes in agricultural practices (Hunter and Leake, 1933). It is clear that wheat breeding and cultivar deployment has had a major effect on the occurrence and intensity of wheat diseases. As some diseases have abated, others have intensified. The risk of disease epidemics will remain high as long as a lack of spaciogenic diversity exists. This is not to say that wheat breeding programs have been unsuccessful. On the contrary, breeding programs have been highly successful within the framework of pure-line development. There are some areas, delimited by climate and geography, that are less of a risk to problems caused by spaciogenic uniformity than are other areas (Lozano and Schwartz, 1981). The lower-risk areas usually have drier environments, such as the western Great Plains of the United States. The rate of disease progress of moisture-requiring pathogens is less in such areas. These areas also have less overwintering of pathogens, particularly the rusts. On the other hand, it is probable that the detrimental effects of other diseases, insect pests, and abiotic stresses could be minimized through planned intra- and interfield diversity (Jensen, 1978; Gill et al., 1984).

The high-risk areas are climatically and edaphically diverse. Incidence and severity of diseases are closely related to the ecology of high-risk areas. In such areas, the relationship between the areal extent of a cultivar and the intensity of diseases peculiar to that cultivar is intimately linked. The space occupied by a genotype is a determinant of disease development as much as disease development is a determinant of the space occupied by the genotype.

The core of the dilemma in breeding disease-resistant cereals is adaptation. Breeders and pathologists cross, select, and develop germplasm that is best adapted (highest yielding; disease resistant) over a number of testing locations and years. Pathogens then adapt to the new germplasm as its use intensifies. The greater the area over which the germplasm is cultivated, the more opportunities the pathogen has to combine overall fitness traits with increased virulence (Watson, 1970). Changes in virulence probably arise continually in pathogen populations. Most of these changes remain undetected because they occur at low frequencies and are often in a genotype that may be less fit. Here, fitness means the per capita rate of increase of an organism (Fisher, 1930), and adaptedness means the inclination of the organism to survive and reproduce (Michod, 1986). Fitness is really the deciding factor in whether or not a virulence change will become pervasive in a population because it tells how much more or less a type will produce over time on a per capita basis (Michod, 1986). Vanderplank (1984) has called pathogen adaptation following widespread production of a cultivar "epidemiologic mutation." Once the pathogen combines the appropriate fitness and virulence traits, its further increase and spread are dependent to a large degree on the spatial extent of its host, given a conducive climate. Thus a biological phenomenon takes on properties of a spatial diffusion problem. Two factors are characteristic of such problems. First, for a pathogen to spread over large areas, it must be disseminated in some way. Carriers include wind, rain, living and dead plant material, and human beings and their equipment. Second, the rate at which a pathogen spreads over space will be influenced by barriers of various kinds. These barriers can be physical (mountains, oceans), biological (nonhosts), climatic (temperature, moisture), political (farm programs, price supports), or cultural (innovative farmers).

The wheat rusts are among the most intensively studied diseases that consistently cause major epidemics. The rusts are highly adaptable to the physical and biological environment, have high reproduction and survival rates, wide genetic variability, and are capable of long-distance dissemination (Johnson, 1978). A recent set of books by Roelfs and Bushnell (1985) has collated the literature on wheat rust epidemics. A major concern of wheat breeders in the first half of this century was the development of wheats resistant to stem rust (Salmon et al., 1953). The breeding method taken to combat the disease was to cross lines containing major-effect genes for resistance into adapted genotypes and produce pure-line cultivars. This led to the cycling of pathogen adaptation and cultivar replacement (Stakeman, 1957; Johnson, 1961).

Along with selection for stem rust resistance, breeders were selecting for wheat genotypes that matured early (Borlaug, 1954). Early maturity added another level of protection to stem rust, because the plants were in a more advanced developmental stage when temperatures became warm enough for rapid stem rust increase. Stem rust causes a greater loss in yield when severities are high at early growth stages (Kirby and Archer, 1927). Thus, by combining genes for stem rust resistance with those for early maturity, epidemics of stem rust have been mitigated since the late 1950s in North America (Roelfs, 1985).

In connection with the epidemic nature of stem rust in the Great Plains of the United States, the following observations were made during the 1985–1986 wheat-growing season. In that year, the wheat acreage along the south Texas Gulf coast doubled, from 24,000 acres to 48,000 acres, mainly due to weak markets in other commodities (Findley, 1985). The predominant cultivars recommended and grown were soft red winter wheats that possessed resistance to leaf rust (caused by *Puccinia recondita*), the most prevalent disease of the area. Fall infections of stem rust were evident by mid-December on wheat planted in early November. When spring temperatures rose, the pathogen proliferated, resulting in the complete destruction of some south Texas wheat fields. As the inoculum migrated northward, some genetic and climatic barriers slowed the rate of spread. However, heavy disease levels were found in Kansas by May, where wheat breeders at Kansas State University were in the process of releasing a new, high-protein wheat called Sumner. Throughout its many years of testing, Sumner was never exposed to such an intensity of stem rust inoculum. As a result, the Kansas wheat breeders recalled the cultivar

(R. G. Sears, personal communication). Because of contiguous wheat cultivation, the lack of spaciogenic diversity, a mild winter, and a "forgotten" disease problem, there were unnecessary losses in yield and germplasm.

Spaciogenic uniformity has contributed to two of the most costly epidemics of leaf rust in North America in modern times. In the 1976–1977 epidemic in Mexico, Jupeteco 73 was planted on 75% of the Sonoran wheat acreage (Dubin and Torres, 1981). To the credit of the Mexican government and CIMMYT, catastrophic losses were averted in Sonora because of timely applications of fungicides. However, losses of more than 40% were realized in some Mexican wheat fields that were not treated (Dubin and Torres, 1981). The cultivar Probrand 812 occupied 63% of the acreage in central and south-central Texas in 1984–1985, when leaf rust caused approximately a 50% loss in yield in the Texas Blacklands and a 28% loss in yield statewide (Marshall, 1988). Both Jupeteco 73 and Probrand 812 were good agronomic and disease-resistant cultivars at the time of their release. It was the growing of these cultivars over large contiguous acreages in high-disease-risk areas that resulted in the epidemics.

Spaciogenic uniformity and wheat intensification have contributed to making some local disease problems regional or even continental. This can be said of septoria tritici blotch, caused by *Mycosphaerella graminicola* (Fuckel) Schroeter (anamorph, *Septoria tritici* Rob. ex Desm.). In this disease, primary infections on seedlings result mainly from propagules produced on debris from the previous years' crop (Shipton et al., 1971). Infections are established on lower leaves and secondary inoculum is later dispersed by splashing rain. The disease tends to develop and the pathogen spreads more rapidly in dense canopies and is favored by long periods of leaf wetness and cool temperatures (Shaner and Finney, 1976).

Septoria tritici blotch had been known to occur on landraces and cultivars in many parts of the world, but rarely had caused severe disease prior to 1960 (Saari and Wilcoxson, 1974). The Mexican semidwarf wheats and the intensified agriculture of the 1960s and 1970s unfortunately served to increase the incidence and severity of the disease in some areas. This was due primarily to the relative susceptibility of the germplasm coupled with conducive climates. High levels of susceptibility were found in Inia 66 in Ethiopia and Tunisia, Siette Cerros in Morocco, and Lerma Rojo and Super X in Turkey (FAO, 1974). The original semidwarf germplasm from Mexico, as a whole, was rather susceptible to *S. tritici* because the germplasm was selected in the near-absence of the pathogen.

Recent evidence suggests that aggressiveness and virulence in *S. tritici* varies with geographic location. Here, "virulence" refers to the ability of the pathogen to produce disease on a cultivar and aggressiveness as the relative amount of damage the pathogen is able to cause. Marshall (1985) found that within the United States, aggressive populations of *S. tritici* were found in north-central California and northern regions of Indiana and Ohio. On a global basis, highly virulent isolates of *S. tritici* were found in Brazil, Mexico, Uruguay, Israel, Tunisia, Turkey, Ethiopia, and Oregon (Eyal et al., 1985). In *S. nodorum,* virulence frequency was high in most isolates from the United States, Canada, and South America, while European isolates were relatively less virulent (Scharen et al., 1985). These works suggest that spaciogenic diversity, in the form of strategic placement of germplasm, could aid in reducing regional spread of the *Septoria* diseases. Previously, Jeger et al., (1981) determined that intrafield spaciogenic diversity could serve to reduce the spread and incidence of septoria nodorum blotch. In a three-year study, Karjalainen (1986) found that disease levels caused by *S. nodorum* were lower in cultivar mixtures than the arithmetic mean of the pure line. However, the mixtures were less effective when disease pressure was high.

Vavilov (1951) cited the case of the pure-line cultivar Svalofs Panzer in Germany in the early 1920s. When the acreage of the cultivar was low, the biotypes of *Puccinia striiformis* virulent on it were absent or strictly limited in their distribution. As the cultivar became more widely grown, biotypes developed that were adapted to the cultivar, producing a serious epidemic of stripe rust in 1923.

In other cereal crops, spaciogenic uniformity has also resulted in numerous disease epidemics. IR-8 and other nitrogen-responsive, semidwarf cultivars developed by IRRI (International Rice Research Institute) were shown to be quite susceptible to bacterial leaf streak and bacterial blight (Singh, 1969; Srivastava, 1967). In India, bacterial blight was unknown in all but two states prior to the introduction of IR-8. Following widespread plantings of the cultivar, the disease became pandemic within two years (Srivastava, 1967). Additionally, the rice semidwarf germplasm reduced the genetic diversity of several Asian breeding programs, as 80% of the crosses made in 1974–1975 carried the same gene for dwarfism (Hargrove, 1979).

Perhaps the most destructive epidemic in terms of amount of plant material lost was the well-documented southern corn leaf blight in the United States in 1970. *Helminthosporium maydis* race T was present and had favorable environmental conditions prior to 1970; however, because of the spaciogenic uniformity of the corn crop, the epidemic occurred (Ullstrup, 1972).

NEW SOLUTIONS

The links between spatial and other aspects of germplasm utilization are only now being forged. Multidisciplinary approaches to cultivar development should be encouraged. Pathologists need to participate more in cultivar release and deployment. Moreover, because plant breeding and plant disease epidemiology have profoundly influenced each other, scientists should be trained in a combined discipline focusing on epidemiological plant breeding. Such trained scientists would be concerned with the breeding *and* deployment in space and time of germplasm engineered to eliminate plant disease epidemics.

It is clear that spaciogenic uniformity has been an important component in the disease epidemics of cereal crops. Pure-line cultivars do not have the ability to evolve or respond to shifts in pathogen populations. Extensive contiguous plantings of pure-line cultivars in disease-prone areas furthers the risk of incurring severe losses due to plant diseases. Improvement is needed in the methods used to breed disease-resistant germplasm and the strategies practiced to deploy it. Spaciogenic considerations in plant breeding and germplasm utilization offer alternatives to the avoidance of risks in cereal production. Schemes to increase diversity will be most effective if their research and implementation is planned and directed toward solving problems.

Pure-Line, Disease-Resistant Germplasm

It would be counterproductive to regress in any way from the great strides and successful achievements that have been realized in the breeding of pure-line, disease-resistant germplasm. However, the effectiveness of currently practiced methods could be increased if viewed as part of a larger program to achieve spaciogenic diversity.

Several of the cereal breeding programs in the world have been pursuing strategies of host plant resistance to direct and stabilize the evolution of pathogen populations. For clarity, the following terms and definitions will be used in this discussion: race-specific resistance is resistance that is effective against some races of the pathogen but not others; general resistance is resistance that does not interact differentially with races of the pathogen. Most cases of race-specific resistance are studied by Mendelian genetics, while most cases of general resistance require study by biometrical means. The pathogen counterpart of host race-specific resistance is virulence, while pathogen aggressiveness is the counterpart of host general resistance. Major-effect genes are those genes that most often correspond with race-specific resistance. Minor genes most often correspond with general resistance.

The accumulation of resistance genes, whether they be of major or minor effect, into breeding lines and germplasm pools has been a widely practiced and effective breeding method (Rajaram and Torres, 1986). This method could be made more effective if greater effort was placed on the identification of the genes being accumulated. A line may contain a single, major-effect gene even though it has a diverse pedigree and has tested resistant over many locations and years. Resistance gene identification must be coupled with identification and frequency information of virulence genes in the pathogen. This combined information allows researchers to monitor and direct pathogen evolution. Much research has been concentrated in this area and should be continued in order to develop superior breeding lines.

An approach that could aid in breeding for general resistance would be to develop a set of "differentials" for the disease in question and to test these at multiple locations and against various pathotypes, in comparison with breeding lines. The differential set should contain a spectrum of lines ranging from the highest level of general resistance to complete susceptibility. Genes for race-specific resistance should not be in any of the lines of the differential set. Coupled with this research should be an aggressiveness survey of the pathogen. Aggressiveness could be tested on a fully susceptible line, as, for example, infection frequency or sporulation capacity. Such an aggressiveness survey, conducted over a long period, would help determine if shifts in aggressiveness occur as cultivars with higher levels of general resistance are planted on larger acreages.

The breeding of germplasm with durable resistance (Johnson, 1978) is a useful concept, but cultivars possessing this type of resistance can be identified only in a historical sense. The durability of resistance can be shown only after wide cultivation of germplasm for a long period under environmental conditions that favor disease. The underlying mechanisms of durable resistance in those cases studied have been shown to be quite diverse (Johnson, 1984). Perhaps the most useful method to breed for durable resistance is to use proven sources of durability as parents in crosses and subsequently select for the resistance traits of the durably resistant parent.

Most of the resistance breeding methods in this brief discussion have aided in the protection of pure-line cultivars from diseases. However, the effectiveness of these methods could be increased if used as part of a diversification and deployment program. The present methods of producing disease-resistant germplasm are totally miscible with methods to increase spaciogenic diversity.

Intrafield Spaciogenic Diversity

Browning (personal communication) has emphasized that one of the major constraints on the use of diversity for intrafield disease management is a paradigmatic one. That is, the underlying basic idea of cultivar uniformity is so fixed that it may actually prevent scientists from researching and understanding the potential benefits of diversity. Clearly, rethinking is needed about the generalized concepts concerning how much uniformity is really needed in a given crop. As Frankel (1950) stated, ". . . the purity concept has not only been carried to unnecessary lengths, but . . . may be altogether inimical to the attainment of highest production."

Even though pure-lines grown over vast areas have been the root of many cereal crop epidemics, some pure-lines are actually somewhat genetically diverse. Pure-line cultivars are usually uniform only for several, highly recognizable traits (Jensen, 1978). The level of actual genetic homozygosity depends on which generation the plants were bulked together for Breeders Seed increase and whether or not any selection had occurred. Some released cultivars are quite variable because they were the result of an unselected F_2 or F_3 population. The variation in such cultivars is usually most apparent when the cultivars are grown in areas remote from their original testing locations. Environmental or biological stresses to which the cultivar had not been exposed may also cause the recognition of previously unrecognized traits. Examples of some heterogeneous cultivars are Clinton oats (Morey, 1949), Scout wheat (Johnson et al., 1965), Siouxland wheat (W. D. Worrall, personal communication), and Harland barley (Jain and Qualset, 1976). One could speculate that a reason some pure-line cultivars have been successfully grown over large areas and long periods was because the cultivars were heterogeneous for disease resistance and thereby had the "buffering" of a population. Research is needed on the potential spaciogenic effects of releasing unselected, early generation germplasm, particularly when the population is the progeny of two- and three-way crosses. Obviously, genetic drift would occur. But, for example, in small grains used for forage, this would not be a concern.

It is clear that plant breeders and pathologists can have a degree of influence over the level of spaciogenic diversity present in agricultural fields. In wheat, it is true that there are pressures exerted by the seed industry and some producers for crop uniformity. However, the seed industry developed and evolved dependent on the pure-line cultivars provided by plant breeders. As Jensen (1978) pointed out, there are inherent problems associated with the preparation and maintenance of cultivar purity in small grains. The appearance of "off-types," even in highly homozygous material, is not unusual. This is so because of meiotic irregularities, low-level outcrossing, and mechanical handling of seed. The roguing of Breeders and Foundation seed fields is a regular, repeated event for most seed producers.

Moreover, reasonable if not total levels of uniformity for some key agronomic traits, such as height and maturity, need not be sacrificed to obtain intrafield spaciogenic diversity. The breeding of multiline cultivars, whether by the mixing of near-isogenic lines (Browning and Frey, 1969), or the CIMMYT "multicross" multilines (Dubin and Rajaram, 1982), serves to increase the spatial difference among like-genotypes in a field. Politowski and Browning (1978) showed that even though multilines use race-specific resistance to protect the crop, the epidemiologic effect was to slow the rate of disease development relative to the susceptible check. Thus although individual oat plants had monogenic resistance to *Puccinia coronata,* it was the epidemiological resistance of the multiline population that effectively slowed the disease. It has been stated best by Browning (1974) and Parlevliet (1981) that monogenic, race-specific resistance should be used only in diverse populations and, preferably, in genotypes with high levels of general resistance.

Another breeding strategy to increase intrafield, spaciogenic diversity is the use of mass breeding or composite crosses. Suneson (1956) proposed the evolutionary-based method of breeding where selected parents of diverse origin are crossed, bulked, and grown as a population for many generations.

These composites would have different genotype distributions dependent on where they were allowed to undergo natural selection (Suneson and Stevens, 1953; Taylor and Atkins, 1954). A similar mass breeding approach to obtaining high levels of general disease resistance and spaciogenic diversity was proffered by Beek (1986), who obtained increased resistance to several small-grain diseases by a bulk-seed selection technique. In Texas, Marshall (1987) has developed an approach to producing *directed landraces* in which germplasm possessing different types of resistance to many diseases is best matched agronomically by means of statistical pattern recognition. These matched lines are hybridized for two to three cycles in the field using male gametocides. The resulting seed is then separated into various lots and each lot is subjected to a different disease pressure. Each harvested lot is selected for high kernel weight/high kernel density seed. Finally, the lots of seed selected are recombined. The result is a directed landrace possessing spaciogenic diversity and reasonable agronomic uniformity.

Some practical and theoretical breeding strategies have been devised to select plants that perform better in mixed than in pure-line culture (Hamblin et al., 1976; Eagles, 1983; Wright, 1985). These mixture-selection schemes seek to develop lines that produce higher yields when grown under competitive conditions and that are adapted to their competitors. Mixture breeding is a rather uncharted area in the cereal crops, and further research is needed to determine the feasibility of the approach.

In addition to breeding strategies designed to increase spaciogenic diversity, a promising deployment strategy that utilizes existing pure-lines is the use of cultivar mixtures (Wolfe, 1985). When the components of a mixture are selected for similar agronomic characteristics, diseases can be managed while yields are maintained and occasionally increased (Wolfe, 1983). The fact that cultivar mixtures use germplasm that is currently available means that spaciogenic diversity can be increased immediately via this method. Cultivar mixtures would allow producers to extend the period of utilization of susceptible germplasm because the susceptibles could be grown with resistant cultivars without the risk of severe yield loss. Because of the interaction between genotype and environment for each cultivar, mixtures have a more stable performance across environments than do pure lines (Wolfe and Barratt, 1980).

Interfield and Interregional Spaciogenic Diversity

Cultivar recommendations are made to a greater or lesser extent by most state agricultural experiment stations and extension services in the United States and by other government agencies throughout the world. Most often, these recommendations are based on yield results over several locations and years. Greater input by pathologists on cultivar recommendations for an area could effectively increase the level of interfield spaciogenic diversity. It is through cultivar recommendations that pathologists can have a significant impact on interfield diversity. The National Institute of Agricultural Botany in the United Kingdom releases information annually on the recommended cultivars for England and Wales in addition to information on how to choose a cultivar, explanations of disease epidemiology, and methods for varietal diversification (NIAB, 1982). Growing four or five genetically different cultivars on a 1000-acre wheat farm in the Great Plains of the United States provides protection against the ravages of a disease epidemic. Based on pedigree and acreage data, there appears to be an increase in diversity among the more recently released red winter wheat cultivars in the United States (Cox et al., 1986). Although this does not necessarily mean that diversity for disease resistance has increased, it does indicate that more choices are available. The risk to the producer is lessened when cultivars that complement each other in regard to risk aversion are grown on a farm. The risk is still less when diverse populations are grown or when genotypes are mixed prior to planting.

On a regional scale, breeders and pathologists have cooperated in using only certain sources of resistance for specific areas (Browning et al., 1969). The breeding approach to the deployment of resistance genes effectively established genetic barriers over and between large geographical areas. Intuitively, this approach works best when few people are involved in the breeding and development of germplasm, such as is the case with oats in the United States. However, when many people from both public and private concerns are involved, such as in wheat breeding, a type of legislative action may be necessary to achieve resistance-gene deployment (McDaniel, 1985).

Cultivar deployment strategies have been developed in India and Cyprus for wheat leaf rust (Reddy and Rao, 1979; Hadjichristodoulou, 1981) and wheat stem rust control (Bahadur et al., 1984). These strategies are based mainly on knowledge of the epidemiology of the diseases, virulence patterns of the pathogens, and resistance genes in the cultivars. Cultivar deployment establishes genetic barriers to the spread of pathogens such that large-scale epidemics can be averted. In Europe, different resistance genes for powdery mildew of barley have been recommended because of spaciogenic uniformity of the winter and spring crops (Slootmaker, et al., 1984). Much more needs to be known about the effects of deploying cultivars to control one disease on other diseases in an area. Clearly, it is better to use several diversification strategies toward all the potential diseases in an area, to minimize production risks.

CONCLUSIONS

It seems reasonable to assume that a change in the basic philosophy that dictates the methods of cereal breeding to include greater levels of spaciogenic diversity could result in the potential elimination of disease epidemics. Genetic crop uniformity over large, disease-prone areas has proven to be a high-risk gamble for both producers and consumers of agricultural products. Crop stability has not been achieved with the widespread cultivation of pure-line cultivars, and it is unlikely that stability will be met should we continue on the same strict course.

At the beginning of this chapter, the question was posed as to whether production levels could be maintained and increased while concurrently extending the useful life of cultivars, providing durable resistance, eliminating epidemics, and reestablishing spaciogenic diversity. If some of the suggested solutions presented here were fairly and adequately tested, it is highly probable that the question could be answered affirmatively.

Methods to increase diversity within and between fields and regions should be incorporated into the current methods of developing and releasing germplasm. Each breeding program should evaluate and research methods to incorporate spaciogenic diversity into resistance breeding techniques and alternative methods of germplasm release and deployment. No single strategy will be appropriate for all breeding programs, nor even for individual objectives within a single program. The most useful strategies to implement diversity will be, in themselves, diverse. The degree of diversity that is needed will vary, but in areas where disease epidemics are a major threat, diversity is more apt to solve the problem.

REFERENCES

ALLARD, R. W., and HANSCHE, P. E. 1964. Some parameters of population variability and their implications in plant breeding. Adv. Agron. 16:281–325.

ATKINS, I. M. 1980. A history of small grain crops in Texas. Tex. Agric. Exp. Stn. Publ. 2M-7-80.

BAHADUR, P., NAGARAJAM, S., and NAYAR, S. K. 1984. Dynamics of stem rust population in India during 1966–1980 and strategy for its management. Indian J. Genet. Plant Breed. 44:190–200.

BEEK, M. A. 1986. Breeding for horizontal resistance to wheat diseases. FAO Plant Prod. Prot. Pap. 69.

BOHRER, V. L. 1972. On the relation of harvest methods to early agriculture in the Near East. Econ. Bot. 26:145–155.

BORLAUG, N. E. 1953. New approach to the breeding of wheat varieties resistant to *Puccinia graminis tritici*. Phytopathology 43:467.

BORLAUG, N. E. 1954. Mexican wheat population and its role in the epidemiology of stem rust in North America. Phytopathology 44:398–404.

BORLAUG, N. E. 1981. Increasing and stabilizing food production. *In* Plant Breeding II (ed. K. J. Frey), pp. 467–492. Iowa State University Press, Ames, Iowa.

BORLAUG, N. E., and GIBLER, J. W. 1953. The use of flexible composite wheat varieties to control the constantly changing stem rust pathogen. Agron. Abstr., p. 81.

BREHAUT, E. 1933. Cato—The Censor on Farming. Columbia University Press, New York.

BROWNING, J. A. 1974. Relevance of knowledge about natural ecosystems to development of pest management programs for agro-ecosystems. Am. Phytopathol. Soc. Proc. 1:191–199.

BROWNING, J. A., and FREY, K. J. 1969. Multiline cultivars as a means of disease control. Annu. Rev. Phytopathol. 7:355–382.

BROWNING, J. A., SIMONS, M. D., FREY, K. J., and MURPHEY, H. C. 1969. Regional deployment for conservation of oat crown rust resistance genes. Iowa Agric. Home Econ. Exp. Stn. Spec. Rep. 64:49–56.

BUDDENHAGEN, I. W. 1983. Agroecosystems, disease resistance, and crop improvement. *In* Challenging Problems in Plant Health (ed. T. Kommedahl and P. H. Williams), pp. 450–460. APS Press, St. Paul, Minn.

BYERLEE, D., and DE POLANCO, E. H. 1982. The rate and sequence of adoption of improved cereal technologies: the case of rainfed barley in the Mexican Altiplano. CIMMYT Work. Pap. 82/4.

CLARK, J. G. D. 1969. World Prehistory: A New Outline. Cambridge University Press, Cambridge.

COX, T. S., MURPHEY, J. P., and RODGERS, D. M. 1986. Changes in genetic diversity in the red winter wheat regions of the United States. Proc. Natl. Acad. Sci. USA 83:5583–5586.

DALRYMPLE, D. G. 1986. Development and spread of high-yielding wheat varieties in developing countries. US AID, Washington, D. C.

DAMANIA, A. B. 1985. Electrophoretic analysis of *Triticum* landraces from Nepal and their use for improved bread-making quality of wheats. Indian J. Genet. Plant Breed. 45:213–218.

DAMANIA, A. B., JACKSON, M. T., and PORCEDDU, E. 1985. Variation in wheat and barley landraces from Nepal and the Yemen Arab Republic. Z. Pflanzendzuecht. 94:13–24.

DESTLER, C. M. 1968. Forward wheat for New England: the correspondence of John Taylor of Carolina with Jeremiah Wadsworth, in 1795. Agric. Hist. 42:201–205.

DORST, J. C. 1957. Adaptation. Euphytica 6:247–254.

DUBIN, H. J., and RAJARAM, S. 1982. The CIMMYT's international approach to breeding disease-resistant wheats. Plant Dis. 66:967–971.

DUBIN, H. J., and TORRES, E. 1981. Causes and consequences of the 1976–77 wheat leaf rust epidemic in northwest Mexico. Annu. Rev. Phytopathol. 19:41–49.

EAGLES, C. F. 1983. Relationship between competitive ability and yielding ability in mixtures and monocultures of *Dactylis glomerata* (L.). Grass Forage Sci. 38:21–24.

EYAL, Z., SCHAREN, A. L., HUFFMAN, M. D., and PRESCOTT, J. M. 1985. Global insights into virulence frequencies of *Mycosphaerella graminicola*. Phytopathology 75:1456–1462.

FINDLEY, D. S. 1985. Texas Small Grain Statistics. Texas Department of Agriculture, Austin, Tex.

FISHER, R. A. 1930. The Genetical Theory of Natural Selection. Oxford University Press, Oxford.

FONTANA, F. 1767. Observations on the rust of grain. Phytopath. Clas. 2.

FOOD AND AGRICULTURAL ORGANIZATION OF THE UNITED NATIONS. 1974. Proc. 4th FAO/Rockefeller Found. Wheat Seminar, Tehran, Iran.

FRANKEL, O. H. 1950. The development and maintainence of superior genetic stocks. Heredity 4:89–102.

FUSSELL, G. E. 1965. Farming Technique from Prehistoric to Modern Times. Pergamon Press, Inc., Elmsford, N. Y.

GILL, K. S., NANDA, G. S., and GURDEV, S. 1984. Stability analysis over seasons and locations of multilines of wheat (*Triticum aestivum* L.). Euphytica 33:489–495.

GOULD, P. R. 1965. Wheat on Kilimanjaro: the perception of choice within game and learning model frameworks. Gen. Syst. 10:157–166.

HADJICHRISTODOULOU, A. 1981. Cereal diseases in Cyprus—control strategies. Cyprus Agric. Res. Inst. Tech. Bull. 38:3–12.

HAMBLIN, J., ROWELL, J. G., and REDDEN, R. 1976. Selection for mixed cropping. Euphytica 25:97–106.

HARGROVE, T. R. 1979. Diffusion and adoption of semidwarf rice cultivars as parents in Asian rice breeding programs. Crop Sci. 19:571–574.

HARLAN, H. V., and MARTINI, M. L. 1929. A composite hybrid mixture. J. Am. Soc. Agron. 21:487–490.

HARLAN, H. V., and MARTINI, M. L. 1938. The effect of natural selection in a mixture of barley varieties. J. Agric. Res. 57:189–199.

HARLAN, J. R. 1972. Genetics of disaster. J. Environ. Qual. 1:212–215.

HARLAN, J. R. 1975. Crops and Man. ASA Press, Madison, Wis.

HUNTER, H., and LEAKE, H. M. 1933. Recent Advances in Agricultural Plant Breeding. P. Blakiston's Son & Co., Philadelphia.

JACOBSEN, T., and ADAMS, R. M. 1958. Salt and silt in ancient Mesopotamian agriculture. Science 128:1251–1258.

JAIN, S. K., and QUALSET, C. O. 1976. New developments in the evaluation and theory of bulk populations. Proc. 3rd Int. Barley Genet. Symp., pp. 739–749.

JASNY, N. 1944. The Wheats of Classical Antiquity. The Johns Hopkins University Press, Baltimore.

JEGER, M. J., JONES, D. G., and GRIFFITHS, E. 1981. Disease progress of non-specialized fungal pathogens in intraspecific mixed stands of cereal cultivars. II. Field experiments. Ann. Appl. Biol. 98:199–210.

JENSEN, N. F. 1952. Intra-varietal diversification in oat breeding. Agron. J. 44:30–34.

JENSEN, N. F. 1978. Population variability in small grains. Agron. J. 70:153–162.

JOHNSON, T. 1961. Man-guided evolution in plant rusts. Science 133:357–362.

JOHNSON, R. 1978. Practical breeding for durable resistance to rust diseases in self-pollinating cereals. Euphytica 27:529–540.

JOHNSON, R. 1984. A critical analysis of durable resistance. Annu. Rev. Phytopathol. 22:309–330.

JOHNSON, V. A., SCHMIDT, J. W., DRIER, A. F., and MATTERN, P. J. 1965. Registration of Scout wheat. Crop Sci. 5:485–486.

KARJALAINEN, R. 1986. Spring wheat mixtures in northern crop production: ability of mixtures to buffer disease development and yield loss caused by *Septoria nodorum.* J. Agric. Sci. Finl. 58:33–42.

KIRBY, R. S., and ARCHER, W. A. 1927. Diseases of cereal and forage crops in the United States in 1926. Plant Dis. Rep. Suppl. 53:110–208.

LOZANO, J. C., and SCHWARTZ, H. F. 1981. Constraints to disease resistance in various food crops grown in Latin America. *In* Proceedings of Symposia of IX International Congress of Plant Protection (ed. T. Kommedahl), pp. 35–38. APS Press, St. Paul, Minn.

LUCE, R. D., and RAIFFA, H. 1958. Games and Decisions: Introduction and Critical Survey. John Wiley & Sons, Inc., New York.

MARSHALL, D. 1985. Geographic distribution and aggressiveness of *Septoria tritici* on wheat in the United States. Phytopathology 75:1319.

MARSHALL, D. 1987. A new strategy for breeding diverse barley germplasm. Tex. Agric. Exp. Stn. Misc. Publ. 3M-7-87.

MARSHALL, D. 1988. Characteristics of the 1984–85 wheat leaf rust epidemic in central Texas. Plant Dis. 72:239–241.

MARSHALL, D., GARDENHIRE, J. H., GILMORE, E. C., MCDANIEL, M. E., and ERICKSON, C. A. 1986. Collin—a new semidwarf hard red winter wheat for the Texas blacklands. Tex. Agric. Exp. Stn. Misc. Publ. 1615.

MARSHALL, D. R. 1977. The advantages and hazards of genetic homogeneity. Ann. N.Y. Acad. Sci. 287:1–20.

MCDANIEL, M. E. 1985. Geographic gene deployment as a disease-resistance strategy. Agron. Abstr., p. 63.

MICHOD, R. E. 1986. On fitness and adaptedness and their role in evolutionary explanation. J. Hist. Biol. 19:289–302.

MOREY, D. D., 1949. The extent and causes of variability in Clinton oats. Iowa Agric. Exp. Stn. Bull. 363.

MUNDT, C. C., and BROWNING, J. A. 1985. Genetic diversity and cereal rust management. *In* The Cereal Rusts, Vol. II (ed. A. P. Roelfs and W. R. Bushnell), pp. 527–560. Academic Press, Inc., Orlando, Fla.

NATIONAL ACADEMY OF SCIENCES. 1972. Genetic vulnerability of major crops.

NATIONAL INSTITUTE OF AGRICULTURAL BOTANY. 1982. Identification and control of cereal diseases.

NEGASSA, M. 1987. Possible new genes for resistance to powdery mildew, septoria glume blotch, and leaf rust of wheat. Plant Breed. 98:37–46.

PARLEVLIET, J. E. 1981. Disease resistance in plants and its consequences for plant breeding. *In* Plant Breeding II (ed. K. J. Frey), pp. 309–364. Iowa State University Press, Ames, Iowa.

POLITOWSKI, K., and BROWNING, J. A. 1978. Tolerance and resistance to plant disease: an epidemiological study. Phytopathology 68:1177–1185.

QUALSET, C. O. 1975. Sampling germplasm in a centre of diversity: an example of disease resistance in Ethiopian barley. *In* Crop Genetic Resources for Today and Tomorrow (ed. H. Frankel and J. G. Hawkes), Cambridge University Press, Cambridge.

RAJARAM, S., and TORRES, E. 1986. An integrated approach to breeding for disease resistance: the CIMMYT wheat experience. *In* Genetic Improvement in Yield of Wheat (ed. E. L. Smith), pp. 55–70. CSSA Spec. Publ. 13.

REDDY, M. S. S., and RAO, M. U. 1979. Resistance genes and their deployment for control of leaf rust of wheat. Indian J. Genet. Plant Breed. 39:359–365.

ROELFS, A. P. 1985. Epidemiology in North America. *In* The Cereal Rusts, Vol. II (ed. A. P. Roelfs and W. R. Bushnell), pp. 404–434. Academic Press, Inc., Orlando, Fla.

ROELFS, A. P., and BUSHNELL, W. R. (Eds.) 1985. The Cereal Rusts, Vols. I and II. Academic Press, Inc., Orlando, Fla.

SAARI, E. E., and WILCOXSON, R. D. 1974. Plant disease situation of high yielding dwarf wheats in Asia and Africa. Annu. Rev. Phytopathol. 12:49–68.

SALMON, S. C., MATHEWS, O. R., and LEUKEL, R. W. 1953. A half century of wheat improvement in the United States. Adv. Agron. 5:3–151.

SCHAREN, A. L., EYAL, Z., HUFFMAN, M. D., and PRESCOTT, J. M. 1985. The distribution and frequency of virulence genes in geographically separated populations of *Leptosphaeria nodorum*. Phytopathology 75:1463–1468.

SHANER, G., and FINNEY, R. E. 1976. Weather and epidemics of *Septoria* leaf blotch of wheat. Phytopathology 66:781–785.

SHIPTON, W. A., BOYD, W. R. J., ROSIELLE, A. A., and SHEARER, B. I. 1971. The common *Septoria* diseases of wheat. Bot. Rev. 37:231–262.

SIMMONDS, N. W. 1962. Variability in crop plants, its use and conservation. Biol. Rev. Camb. Philos. Soc. 37:422–465.

SINGH, K. G. 1969. Bacterial leaf streak in west Malaysia. FAO Plant Prot. Bull. 17:64–66.

SLOOTMAKER, L. A. J., FISCHBECK, G., SCHWARZBACH, E., and WOLFE, M. S. 1984. Gene deployment for mildew resistance of barley in Europe. Vortr. Pflanzen. 6:72–84.

SRIVASTAVA, D. N. 1967. Epidemiology and control of bacterial blight of rice in India. Proc. Symp. Rice Dis. Control, pp. 11–18. Japan, Agric. For. and Fish. Res. Council.

STAKEMAN, E. C. 1957. Problems in preventing plant disease epidemics. Am. J. Bot. 44:259–267.

SUNESON, C. A. 1956. An evolutionary plant breeding method. Agron. J. 48:188–191.

SUNESON, C. A., and STEVENS, H. 1953. Studies with bulked hybrid populations of barley. USDA Tech. Bull. 1067.

TAYLOR, L. H., and ATKINS, R. E. 1954. Effects of natural selection in segregating populations upon bulk populations of barley. Iowa State Coll. J. Sci. 29:147–162.

TOZZETTI, G. T. 1767. True nature, causes and sad effects of the rust, the bunt, the smut, and other maladies of wheat, and oats in the field. Phytopath. Clas. 9

UCKO, P. J., and DIMBLEBY, G. W. 1969. The Domestication and Exploitation of Plants and Animals. Gerald Duckworth & Co. Ltd., London.

ULLSTRUP, A. J. 1972. The impacts of the southern corn leaf blight epidemics of 1970–71. Annu. Rev. Phytopathol. 10:37–50.

VANDERPLANK, J. E. 1984. Disease Resistance in Plants, 2nd ed. Academic Press, Inc., Orlando, Fla.

VAVILOV, N. I. 1951. The Origin, Variation, Immunity and Breeding of Cultivated Plants. The Ronald Press Company, New York.

WAHL, I., ANIKSTER, Y., MANISTERSKI, J., and SEGAL, A. 1985. Evolution at the center of origin. *In* The Cereal Rusts, Vol I (ed. A. P. Roelfs and W. R. Bushnell), pp. 61–95. Academic Press, Inc., Orlando, Fla.

WATSON, I. A. 1970. Changes in virulence and population shifts in plant pathogens. Annu. Rev. Phytopathol. 8:209–230.

WITCOMBE, J. R., and GILANI, M. M. 1979. Variation in cereals from the Himalayas and the optimum strategy for sampling plant germplasm. J. Appl. Ecol. 16:633–640.

WOLFE, M. S. 1983. Genetic strategies and their value in disease control. *In* Challenging problems in plant health (ed. T. Kommedahl and P. H. Williams), pp. 461–473. APS Press, St. Paul, Minn.

WOLFE, M. S. 1985. The current status and prospects of multiline cultivars and variety mixtures for disease resistance. Annu. Rev. Phytopathol. 23:251–273.

WOLFE, M. S., and BARRATT, J. A. 1980. Can we lead the pathogen astray? Plant Dis. 64:148–155.

WRIGHT, A. J. 1985. Selection for improved yield in inter-specific mixtures or intercrops. Theor. Appl. Genet. 69:399–407.

10
Yield Variability and the Impact on Local and Regional Estimates of Yield Reductions Caused by Disease

R. E. Gaunt and M. J. Robertson

INTRODUCTION

Plant pathologists have developed a variety of methods for measuring the amount of disease on single plants and in crops. The methods are based on the frequency of pathogens and the intensity of the diseases they cause, as reviewed by Teng (1987) and James and Teng (1979). The information has been used for several purposes, including the analysis of epidemics, the assessment of disease resistance, fungicide efficacy, and yield loss. Recently, there has been an increased awareness of the importance of the spatial patterns of disease within single plants, crops, and regions, as reviewed in other chapters of this book. Average values for incidence or severity, useful though they may be for some purposes, are of limited value for a full understanding of the development of disease. There has been a similar interest in the variation in yield among individual plants and crops by crop physiologists and agronomists concerned with explaining the effects of the environment or management on yield.

The traditional approach to the estimation of regional crop loss (Teng and Shane, 1984) has been to define a disease severity:yield loss relationship,

Note: Department of Agricultural Microbiology, Lincoln University College of Agriculture, Canterbury, New Zealand. Department of Agriculture, University of Queensland, St. Lucia, Queensland 4067, Australia.

to measure the intensity of disease within the region by survey methods (Main and Proctor, 1980), and from these, to calculate the regional loss. The relationship usually is derived by empirical methods, often by regression analysis, based on specific experimental work designed for the purpose (Sah and MacKenzie, 1987; Shane and Teng, 1987). Yield loss is commonly expressed as a proportion of yield in the absence of disease, although some models express the loss in absolute units (MacKenzie and King, 1980). Disease is usually measured, either directly or indirectly, as disease severity on a proportional scale. Plants and crops are selected for the measurement of disease as being representative of the production system, using recommended sampling procedures (Cochran, 1977). The accuracy of regional loss estimates derived by this approach is determined by the validity of the models and the survey information. The approach assumes that, on average, there is a constant relationship between disease intensity and yield loss which is not affected significantly by local or regional variations in climatic, edaphic, and management factors that influence both pathogen and crop development. The assumption will be valid only if the model was derived from data that encompassed the full range of variation present in the production system. Proportional-loss models assume that there is less absolute loss per unit of disease severity in crops with low yield potentials compared with crops with high yield potentials. On the other hand, absolute loss models assume that the actual loss per unit of severity is the same. Both assumptions are simplistic and may often be incorrect in the field.

An alternative approach is to derive mechanistic models of crop production, including the effect of disease constraints (Boote et al., 1983). Yield loss is the result of complex interactions between disease development and the growth and development of plants. The response of single plants, and therefore the crop, to the presence of disease is dependent on plant factors which determine plant sensitivity, as well as the duration, timing, and severity of disease (Lim and Gaunt, 1986; Thomson and Gaunt, 1986). A mechanistic model of plant growth is the ideal medium for definition of these factors, based on an understanding of host metabolism and partitioning of resources. Process-level models, based on such plant processes as photosynthesis, respiration, and water uptake, have not been successful for the prediction of yield over a wide range of environments, although they have been extremely useful in investigations of yield limitation. Crop-level models predict yield more successfully, but they include a considerable degree of empiricism within a mechanistic framework. The advantages of the mechanistic approach, based on an understanding of yield physiology, have been reviewed recently from the perspective of disease management (Gaunt, 1987). There may also be some advantages in this approach for crop-loss estimation in areas where the production environment is heterogeneous and where the resultant regional variation in yield, even in the absence of disease, is large.

Many investigators have shown that yield does vary markedly both locally and regionally, usually in relation to one or a few factors. The landmark investigations by Rothamsted scientists on multiple factors influencing yield (Prew et al., 1983, 1985) highlighted the complex interactions that occur and emphasized the major influence of disease and pests in the British cereal production system. In three seasons maximum yields were achieved under different treatment combinations, which reflected the factors that were most limiting in those seasons. The incorporation of an understanding, and prediction, of those limiting factors in disease severity:yield loss models is a challenge to present-day pathologists and physiologists. In this chapter we review the sources and amount of yield variation in crops in the absence of disease and pest constraints and then discuss the relevance of this variation to the response of crops when the pest and disease constraints are present. Spatial variation has been interpreted in a broad sense to include all factors that may contribute to differences in yield except disease and pest constraints. Diseases and pests are themselves subject to variation as discussed in Chapter 8, and they may contribute further to yield spatial variation.

SOURCES OF VARIATION

Variation is caused by differences in edaphic, weather, management, and genetic factors. These factors interact to control the processes of crop growth and development. The major factors discussed are soil nutrient availability, drought and waterlogging, temperature, and incident light.

Nutrient Availability

Most production systems are subject to limitations in nutrient uptake, and the variation both locally and regionally often has a greater effect on yield than that of any other of the factors experienced. The availability of the major nutrients (i.e., nitrogen, phosphorus, potassium, calcium, and magnesium) is influenced by complex interactions between soil type, climate (especially rainfall), and previous management (especially crop sequence). The effect on yield may be large, especially on land recently cleared of native vegetation for crop production (Trangmaar et al., 1987). The greatest variation occurs between soil types, followed by between-field and within-field variation. Small to medium-sized fields are likely to contain only one soil type, and the variation found within the field is often no more than twice the variation found within a 1-m^2 area anywhere in the field (Beckett and Webster, 1971). Despite the relative importance of nutrient availability, as evidenced by the fertilizer inputs in many developed agricultural systems, there has been little study of the within-field yield variability caused by variation in nutrient availability; most work has concentrated on the response to fertilizer inputs in specific situations.

Nitrogen has received the greatest attention as a constraint to yield, and therefore as a source of variability. Commonly, but not exclusively, it has been shown to be the major limiting factor to production. The uptake of nitrogen by a crop depends on the amount of fertilizer and soil nitrogen available, and the efficiency with which it is recovered. The latter may vary between 15 and 90%, although recoveries usually range from 40 to 70% in most systems (Dowdell, 1982). Soil moisture is a major factor in recovery, because of an effect on the processes of leaching and denitrification. This effect may be so extreme that yields are depressed by above-normal midwinter rainfall (Fisher, 1924). Soil reserves of nitrogen may vary greatly, as shown by the range of 30 to 150 kg N/ha in spring under winter wheat crops in the Netherlands (Loomis, 1983). This variation, mostly between rather than within fields, as shown in Table 1, is influenced by soil type and the preceding crop. For example, the presence of a leguminous crop in a rotation every two years may double cereal yields (Loomis, 1983). The response to increased nitrogen uptake when nitrogen is limiting is predictable [(e.g., 50 to 80 kg of grain per kilogram of nitrogen in cereals (Van Keulen, 1986)]. The variation in this response is attributable to variation in grain:straw ratios related to growing conditions and cultivar characteristics. However, the response to fertilizer inputs is much less easy to predict, because it is difficult to assess soil nitrogen reserves and their availability. This is illustrated by the results of 44 experiments on the response of winter wheat to spring nitrogen applications at five rates at different locations (Batey, 1976). At more than half the sites the advisory services recommendation for nitrogen application differed from the optimum by more than 40 kg N/ha. Similarly, at one-fourth of the sites the yield at the recommended rates differed by more than 0.6 ton/ha from that at the optimum rate.

There has been less work on the variation in yield caused by nutrients other than nitrogen, but a similar relationship occurs between yield and uptake of other nutrients (Van Keulen, 1986). However, the availability of each nutrient is likely to be affected by different aspects of the environment. For example, variation in phosphorus availability is influenced in some systems more by the previous crops and added fertilizers than by the soil type (McCollum and McCabb, 1954). Recent interest in the role of endomycorrhizal fungi in the uptake of phosphorus has highlighted the variation that may occur in

TABLE 1 Topsoil Coefficients of Variation (%) for Nitrogen Content for Increasing Sample Areas

	Size of Sample Area				
	Within Field				
	1 m	0.01 ha	Total	Several Fields	Between-Field Component
% CV	10–20	10–20	25–30	30–55	25

Source: Bennett et al. (1980).

availability of this nutrient, but there is little information available on small-scale variations within fields.

Soil Moisture Status

Drought is a major factor in the variation in yield in many production systems. In the semi-arid tropics, where temperature is not a limiting factor, the length of the growing season is determined by the precipitation pattern (Reddy, 1984). This pattern may change, thus causing major variation in yield potential. For example, in West Africa the annual rainfall varies from 1250 to 250 mm from latitude 10°N to 17°N, respectively. The growing season ranges from 200 to 25 days over the same range of latitude (Konato, 1984).

In less extreme situations, drought does not limit growing-season length but may cause major variation in the yield attained within the growing season. The degree of drought is related to the seasonal balance of rainfall (or other precipitation) and total evapotranspiration from the canopy and soil surface. Before discussing these factors, it is useful to define a measure of drought. The limiting potential soil moisture deficit (Dl) is the amount of water (millimeters) evaporated from the soil profile to the point where crop growth effectively ceases (Penman, 1962). The value of Dl depends on both crop and soil type, and is large for deep-rooted crops and for soils with a large available water-holding capacity. For a given climate, crops with a large Dl are less likely to be subject to drought stress than are crops with a small Dl. Crop yield is reduced below its potential whenever Dl is exceeded during the season; the size of the yield reduction is proportional to the length of time that the Dl value is exceeded.

Evapotranspiration does not vary markedly over relatively long distances, because it is determined mostly by large-scale weather parameters such as net radiation and atmospheric vapor pressure deficit. For example, crop evaporation in most of Britain has a rate of 2.5 to 3 mm/day during the summer and annual potential evaporation ranges between 450 and 500 mm, with little variation from year to year (Smith, 1976). Similarly, the total evapotranspiration during May–July over the 1500-km breadth of the Canadian Great Plains varies from 350 to 400 mm (Williams, 1971).

Variation in evapotranspiration does occur, however, at the local level. In hilly areas there are large differences in evapotranspiration between slopes with different aspect because of the influence on net radiation. For example, north- and south-facing slopes at Ballantrae, New Zealand, had average net radiation values of 6.7 and 4.5, respectively (Lambert and Roberts, 1976). The values for potential evapotranspiration were 3.2 and 1.57 mm/day, respectively. Large areas of uniform vegetation will have very little variation in evapotranspiration. The edges of these areas, and the whole of smaller areas, may be influenced markedly by shelter and the management of adjacent fields. Shelter reduces wind speeds, which results in reduced evapotranspiration, often causing increases in yield extending for a distance 8 to 10 times the height of the barrier. The effect may be especially marked in dry seasons or areas and in soils with low water-holding capacity (Marshall, 1967). Evapotranspiration may also be increased locally by the presence of a dry fallow area upwind of the crop, especially if the crop is irrigated. This is caused by the dry and warm air associated with movement over the fallow area (Davenport and Hudson, 1967), and may cause considerable yield reduction on the margins of the crop.

Precipitation, in contrast to evapotranspiration, is more variable in space. Rainfall varies markedly over small areas of uniform topography and is also influenced by altitude. Snow is subject to additional influences related to shelter (Marshall, 1967) and wind speed. In regions where Dl is exceeded in most seasons, rainfall is often directly correlated with yield variation and may be a dominant factor (Fig. 1). For example, wheat yields on the Canadian Great Plains may be grouped into eight crop districts based on characteristics of preseason conserved soil moisture and summer rainfall. The effect of precipitation in these areas is so dominant that the soil type and variability in seasonal evapotranspiration can be ignored (Williams, 1971). In those regions where Dl is not often exceeded, precipitation has little effect on yield variability (Monteith, 1981), except where flooding causes variation associated with waterlogging.

Variation in the depth of the soil profile available to plant roots may contribute to yield variation, especially in regions where Dl is often exceeded. This is particularly the case where the depth is less than the potential rooting depth. Soils on alluvial floodplains are characterized by short-range variability in texture and depth (Karageorgis et al., 1984). For example, Bennett et al. (1980) showed that depth within a field ranged between 150 and 900 mm of

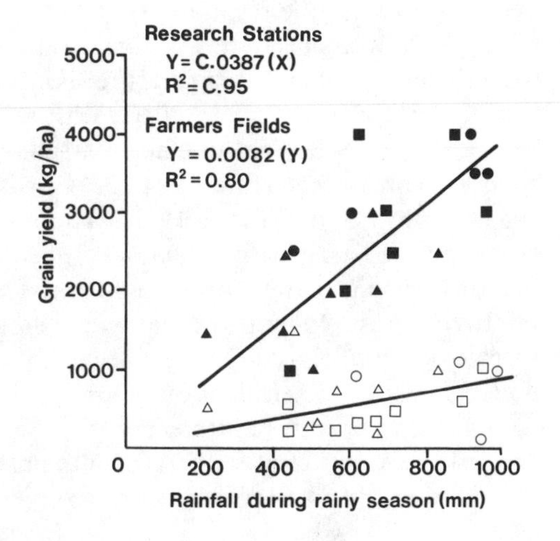

Figure 1 Relationship between rainfall and yield of maize (circle), sorghum (square), and millet (triangle) grown at research stations (filled) and farmers' fields (open) in India. [From Sivakumar, M. V. K., Huda, A. K. S., and Virmani, S. M. 1984. Physical environment of sorghum- and millet-growing areas in south Asia. Page 64 *in* Agrometeorology of sorghum and millet in the semi-arid tropics: proceedings of the International Symposium, 15–20 Nov 1982, ICRISAT Center, India. Pantancheru, A.P. 502 324, India: International Crops Research Institute for the Semi-Arid Tropics. (CP 138)]

fine-textured material overlying gravel. In a dry season, when evapotranspiration exceeded rainfall by 430 mm, the yield of spring-sown barley, with a potential rooting depth of 1 m, varied within the field from 2.8 to 6.8 tons/ha. In a wetter season (270 mm deficit) the yield varied less: from 2.3 to 3.2 tons/ha. An additional factor of major importance when considering drought is the potential rooting depth of crops, which influences the value of Dl and therefore the likelihood of drought being experienced in a particular climate. Monteith (1981) estimated that rainfall variation caused a 9% yield variation from year to year in potatoes in Britain on heavy soils, but only a 2% variation in deeper-rooted spring cereals.

Potential yield for a wide range of crop species in the absence of drought is decreased by 0.1 to 0.3% for each millimeter of water by which Dl is exceeded (French and Legg, 1979; Penman, 1962, 1970; Wilson, 1985). Thus the yield of a wheat crop with a 6-ton/ha potential yield will be reduced by 120 to 360 kg/ha by a 20-mm increase in the potential soil moisture deficit in excess of Dl.

In summary, the relative importance of factors that cause yield variability because of drought depends on the spatial scale being considered. Variation in soil depth is especially important in causing within- and between-field variation, especially in dry areas. On the regional scale, precipitation is the major source of variation and it is only on a larger scale that variation in evapotranspiration has to be considered. The degree to which Dl is exceeded during the season indicates the severity of drought.

Temperature

Temperature is often extremely variable both in time and space within and between crops. Except in extreme conditions, plants integrate temporal changes in temperature during a day and it is therefore useful for most purposes to consider the response to daily mean temperature (Arnold and Monteith, 1974). It has been shown in controlled environments that the rate of photosynthesis is related to temperature. For temperate C3 crops there is a linear increase over the range 5 to 20°C, with an approximate response of 4% per °C. (Ludlow, 1983; Woledge and Parsons, 1986). For C4 crops, a linear response occurs over the range 10 to 35°C, with a 20% increase per °C (Ludlow, 1983; Ludlow and Wilson, 1971). The implication for variation is that C3 crops are far less sensitive than C4 crops to small variations in temperature. For this reason, many workers ignore the effect of temperature on C3 crop photosynthesis. Nonetheless, daily temperature is usually included in agrometeorological models of crop yield because it does influence two additional components of yield determination.

First, temperature influences the rate of crop establishment and canopy expansion (Angus et al., 1980). Small temperature changes may induce large responses in the time taken to full canopy light interception, and this effect is reflected in final yield in many crops. For example, a decrease in mean daily temperature of 1°C during April to July in Britain was predicted to reduce seasonal light interception of sugar beet crops by 10%, resulting in a proportional yield reduction (Scott and Jaggard, 1978). Similarly, the yield of forage oats was influenced markedly by temperature during early season growth in New Zealand (Hughes et al., 1984). At a northern warm site the time to full canopy cover was 50 days, whereas at a more southern site it was 150 days. As a consequence, incident radiation was utilized more effectively at the warmer site, and after 200 days the yields were 17.5 and 8 tons/ha at the warm and cool sites, respectively.

Temperature also affects the length of the growing season in many crops. In vegetative crops growth occurs only during that time of the year when mean temperatures are above the base temperature for leaf growth. The period of vegetative growth may also be influenced by the induction of reproductive growth in some crops, thus curtailing the production of harvestable yield (e.g., some grasses). Vegetative yield of sugar beet is related to the total accumulated day-degrees during the period of active growth (Scott et al., 1973), which is influenced markedly by temperature at different latitudes (Loomis, 1983). Annual pasture yield in New Zealand hill country varies with altitude by approximately 90 kg dry matter per 100 m of altitude, mostly because of the change in growing period (Cossens, 1983).

The effect of temperature on phenological development in crops harvested for their reproductive component is also an important determinant of final yield. The time from sowing to maturity is reduced by increasing temperature, because of a faster rate of development (Ellis and Kirby, 1980). Dry matter accumulation is therefore reduced because of the shorter growth period (Monteith and Scott, 1982). Thus temperature has opposing effects on yield in reproductive crops: during early vegetative growth, increased temperatures increase production by reducing the period to full light interception, but increased temperatures overall reduce the period to maturity, thus reducing potential yield (Ellis and Kirby, 1980). In many annual reproductive crops yield is negatively correlated with temperature, implying that the dominant effect of temperature is on the duration of crop growth. For example, wheat yields in Britain decrease by 5% for each 1°C increase in mean growing-season temperature (Fig. 2).

It is important to recognize that the degree of yield sensitivity to temperature is related to the base temperature for development of the crop species. Each species has a characteristic base temperature, which is greater for tropical than for temperate species. If a temperate species with a base temperature close to 0°C (e.g., wheat) is subject to a mean temperature increase from 15°C to 16°C, development will be 7% faster and growth duration will be reduced proportionately. In contrast to this, a species with a base temperature of 10°C (e.g., maize) will respond with a 20% increase in development rate and a correspondingly greater yield response to the same 1°C change in temperature

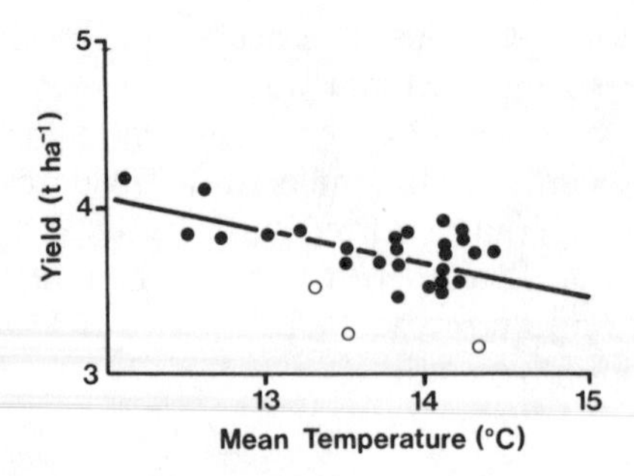

Figure 2 Mean yield of wheat in England (1956–1970) as a function of mean temperature for May, June, and July for the period 1941–1970. Line of best fit has slope of –5% at 14°C. (From Church and Austin, 1983.)

(Cooper, 1979). Thus the relative importance of temperature variation will be dependent on the crop species as well as the prevailing temperature of the production area.

The temperature effects described above influence yield and therefore cause spatial variation. Shelter at the edge of fields causes changes in mean daily temperature, although these changes are usually small [e.g., a 2°C increase in maximum temperature (Marshall, 1967)]. It is likely that temperature changes may account for only 10% of the increase of yield associated with shelter. The major effect of temperature on yield of both vegetative and reproductive crops is seen on the regional and global scales, related to changes in altitude and latitude. Temperature effects will cause yield variation when crops are sown at different times within the same area. Thus adjacent fields may have very different yield potentials if sown at markedly different times, especially in determinate reproductive crops such as cereals. Barley crops sown in April in New Zealand had greater yields than those sown in July or September, related to effects on the crop growth period.

Light

Leaves and single plants respond to variations in light energy inputs by different photosynthesis rates (Goudriaan and van Laar, 1978). However, the growth of C3 crops is not very sensitive to moderate variations in radiation. In Western Europe daily light integrals of between 17 and 20 MJ are associated with growth rates of crops with full canopy cover of between 150 and 200 kg dry matter per hectare per day for a wide range of vegetable and arable species unaffected by drought or disease (Monteith and Scott, 1982). This comparative lack of response in C3 crops was explained by Monteith (1981) as the effect of light saturation. He suggested that year-to-year variation of crop photosynthesis during the summer in Britain was about 3% for a 10% variation in net radiation. However, C3 crops will respond to changes in radiation in the spring and autumn in temperate areas when light saturation is infrequent. For example, the yields of spring barley in Britain respond to above-average light receipts in May (Jones, 1979). The lack of sensitivity in C3 plants is demonstrated by the small response in potential yield to changing radiation inputs over a range of latitudes of 10 to 40° (Fig. 3). The yield potential decreases with latitude changes higher than 45° because light saturation occurs less frequently (Loomis, 1983). In contrast, C4 plants respond to a wider range of light intensity and total radiant energy receipts, because light saturation thresholds are higher than in C3 plants. In high-light environments, C4 plants may convert 5 to 6% of the light energy compared with 3 to 4% for C3 plants, assuming that other factors, such as temperature, are not limiting (Cooper, 1975). Between latitudes 10°N and 20°N, the potential yield of C4 plants may change by up to 30% in response to changing radiation (Jones, 1979).

In practice, radiant energy inputs are responsible for yield variation in few situations, because other factors are usually more limiting. In the tropics, where C4 crops may be expected to show variation in response to light receipts, yields are often limited by drought and nutrients in low-input farming systems.

RESOLUTION AND MAGNITUDE OF VARIATION

The soil and weather factors and management factors described above may operate at one or more scales of resolution from individual plant parts to cultivated areas. The amount of within-plant variation is dependent primarily on plant species. In those crops in which seed is the harvested component, the mean seed weight is relatively stable. The major variation is associated mostly with seed number per plant except in extreme circumstances. Similarly, fruit weight is often very stable in those production systems that are bred and managed to meet a consumer or market demand for uniform size. The variation in seed or fruit number between individual plant parts depends largely on plant architecture. For example, the number of grain per spikelet within an ear or per tiller within a cereal plant may be very variable, and seed number per pod may be the major source of variation in some legumes.

Variation between individual plants is dependent partly on the type of

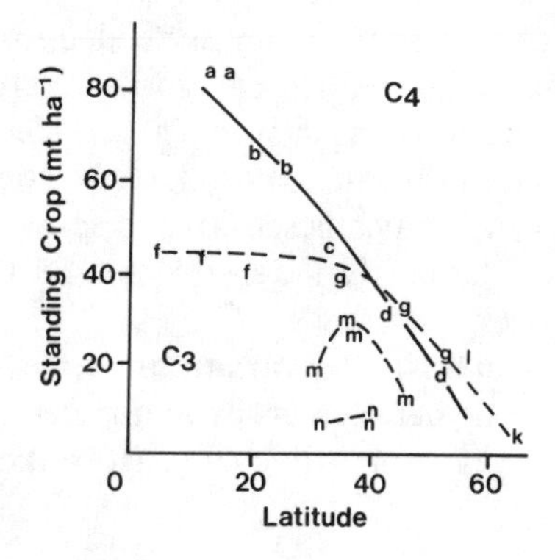

Figure 3 Relation between record annual production of several C3 and C4 crop species and latitude. Trend lines are fitted by eye to the following crops: a, napier grass; b, sugarcane; c, forage sorghum; d, maize; f, cassaya; g, sugar beet; k, wheat; l, ryegrass; m, alfalfa; n, soybean. (From Loomis, 1983.)

plant population. In natural populations with low densities and high species diversity, each individual may vary markedly in productivity. As density increases the individual is subordinated to the population as a whole, but productivity per unit area may still be dominated by a few large individuals. The different age structure of the population is one factor that maintains heterogeneity, such that the commonest plant size is not the average or the most frequent (Harper, 1977). In single-aged monocultures, variation is likely to be lower than in natural populations, but variation does occur, as seen in maize crops (Yoda et al., 1957). The yield of an individual plant was shown to be negatively correlated with the adjacent plants and positively correlated with that of plants two and four positions distant. There is little information available in the literature on plant-to-plant variability, but this variation may be an important factor to consider, especially as disease is often not distributed randomly between plants. Studies based on single plants or tillers (King, 1976) with a range of disease severity on individual plants could provide detailed information on the range in yield of healthy plants as well as the response of individual plants to the presence of disease.

Within fields, there may be considerable yield variation between small areas caused by changes in soil type. Irregularities in the application of fertilizers and water by irrigation may also cause variation. Between larger areas within fields, yield variation may be caused by shelter, topography, and previous rotation history, in addition to the effects of soil type and depth. Variations in sowing density, associated with drilling inaccuracies, may contribute to yield variation. However, in commercial situations, crops are usually sown at densities that do not respond markedly to small variations.

There are often variations in yield between different fields in local areas, even though these areas may appear to be uniform in most respects. Microclimatic effects and management decisions with respect to genetic and other inputs may result in a wide range of yield. Yield structure associated with different growth and development patterns may also vary markedly between neighboring crops, as shown in the contribution of yield components of different wheat cultivars with similar yields (Austin et al., 1980). Further variation in yield between different cultivars may be associated with different harvest indexes, even though the biomass production may be similar. Apparently similar fields may contain different cultivars, crops sown at different times of the year, different cropping history, and different irrigation and fertilizer practices. This variation will be correlated to some degree with spatial variation in disease intensity, which is sensitive to some of the same factors. However, the variation will be partly independent, and therefore important in relation to crop sensitivity to diseases.

There will be similar variations because of climatical management factors on a larger scale between crops in different regions, although the relative importance of the factors will change. It is likely that temperature, drought, and possibly light may contribute markedly to variations in potential yield and yield structure on this larger scale.

The relative importance of factors contributing to yield variation at different resolution is not well documented, but we have summarized our current views in Table 2. Church and Austin (1983) summarized the amount of total yield variation in wheat in England and Wales, based on commercial farming systems and experimental investigations (Table 3). They also compared these variations with yield responses to experimental treatments in crops under accepted good management. Of 14 estimates of main factor effects on yield, three were negligible, seven were 6% or less, but four factors had average effects from 9 to 16% over all levels of other factors. It is interesting to note that three of these related to pest or disease control.

SOURCES OF DATA

There have been very few specific investigations of spatial yield recently, or of spatial variation of the contributing factors. This may be attributable in part to the successful partitioning of variation in randomized block designs developed by Fisher in the 1920s. However, uniformity trials were carried out by many workers in earlier investigations, often as a preliminary to determining optimum plot size and shape (Mercer and Hall, 1911). These data have been listed by Cochran (1937) and summarized by Fairfield Smith (1938). The latter derived an empirical relationship between variance in yield and plot size and concluded that whereas long narrow plots may on average be less variable than square plots of the same size, the effect was often reversed depending on the orientation of the plots. He also showed that wheat yields varied from 2.4 to 4.3 tons/ha, which he attributed solely to variations in soil fertility. Similarly,

TABLE 2 Factors that Cause Spatial Yield Variation in the Absence of Disease and Pest Constraints

Scale	Factor[a]
Adjacent plants	Sowing density, genetic diversity
Within fields	
<10 m	Soil fertility, depth, and texture
>10 m	Soil type, shelter, edge effects
Between fields	Management, soil type, topography
Between regions	Precipitation, temperature, evapotranspiration, soil type
Between countries	Temperature, precipitation, evapotranspiration, incident radiation

[a]Factors are listed in decreasing order of importance within each scale.

TABLE 3 Indicators of Variability of Wheat Grain Yields per Hectare

	Standard Deviation (%)
Between replicate plots	4–9
Between fields on the same farm	15
Between average individual farm yields	20
Between individual fields in country	25

Source: Church and Austin (1983).

Mercer and Hall (1911) reported a range of ±30% relative to the mean yield of wheat. Harris and Scofield (1928) indicated that plot yields were positively correlated from season to season, implying a permanence in observed variation attributed to soil fertility. Further analysis of the Mercer and Hall observations (McBratney and Webster, 1981) indicated that there was a periodicity in both grain and straw yield not detected in the original paper. This was attributed to an early ridge and furrow system used some time before the observations, again indicating some degree of permanence. Recently, there has been a renewed interest in spatial soil variability, although this has not usually been related to yield variability (Beckett and Webster, 1971; McBratney, 1985).

Information is also available in the literature and in original data files from several other sources. For example, data files may be such that a comparison of individual replicate values, and sample unit values within replicates, may be analyzed for variation on the assumption that analysis error is minimal. This assumption is probably valid for measurements of plant dry matter and yield. Similarly, the full disease and pest control plots in evaluations of chemical efficacy, and in yield loss studies, may be a further source of information. Reference back to original data may provide considerable information without the need for further experimentation.

There have been many reports on the relative yield of different cultivars at multiple locations in several countries carried out as the basis for cultivar recommendations (Mycroft, 1983). Similarly, fertilizer evaluations are often carried out at several locations. In both cases the trials are often conducted with full disease and pest control, thus eliminating that source of variation. The data may be used to analyze locational variation, as reported by Tonkin and Silvey (1982). They reported that the within-trial plot errors assessed as the coefficient of variation were generally low (0.2 to 6.7%) both in the presence and absence of fungicide treatment, but that some cereal crops were more likely to have high variances (e.g., autumn-sown barley). A further source of information may be found in the individual yields of healthy plants in single tiller/plant methods of investigation of disease:yield loss relationships. The regression models in these investigations usually have high variance and account for only a small proportion of the total variation in yield, suggesting that factors other than disease are causing considerable spatial yield variation.

For example, the r^2 statistics for studies of wheat and barley grown in Scotland (Richardson et al., 1976) were less than 0.3, and often less than 0.15.

IMPACT OF YIELD VARIATION ON PRODUCTION CONSTRAINTS

The yield variations discussed above are relevant to the estimation of yield loss, although the degree will depend on the resolution of the estimates. National and global estimates of crop loss for economic purposes, such as marketing and the establishment of research priorities, will be least sensitive to spatial variations in yield, provided that the original models were derived from data that included spatial yield variation. Robust models from data derived from investigations in a range of climates and management practices will be good predictors provided that the original models are not used outside the intended range of inputs. Thus models should be used in new climatic zones or in new management systems with extreme caution until they are validated. Major shifts over time in management practices, such as cultivar types and tillage systems, may render existing models invalid, especially if the changes are associated with changes in mean yield.

The relevance of spatial yield variation increases as the resolution increases, and in the case of individual fields the majority of empirical models are likely to be invalid. In the remainder of this section we examine the implications of a knowledge of spatial yield variation for the development of better models for local and individual field yield loss estimation.

One of the major difficulties in deriving meaningful relationships between production constraints and yield is the identification of the degree of stress. Thus we have shown above that in some cases factors such as temperature and light are limiting, whereas under different conditions these factors have no effect on yield. Disease as a constraint is no different from the abiotic factors and should therefore be analyzed by similar methods. No measure of disease incidence or severity can substitute effectively for the measurement of stress, although there will be some correlation. Unfortunately the degree of correlation is very variable. Stress is a factor determined not only by the amount of the constraining factor present, but also by the degree of sensitivity of the host or host part. The effect of an additional disease constraint, for example, will be dependent partly on the position of the infection on the plant relative to the yield function of that plant part. Thus a small disease lesion on a pod, or leaf subtending a pod, may have the same effect on yield as a much larger lesion on a leaf in the non-reproductive zone of a legume, even if the larger lesion is located in the upper part of the canopy where light interception is high (Griffiths and Amin, 1977). As such, stress will be associated closely with variation in crop growth and development.

There has been major progress in one area relevant to yield variability—

the definition of disease thresholds. Although more familiar in relation to populations of insect pests, disease thresholds have been defined for a number of diseases (Cole and Gaunt, 1987; Hills and Worker, 1983; Zadoks, 1985, 1987) and have been incorporated into management models such as EPIPRE (Rabbinge and Rijsdijk, 1984; Zadoks, 1984). In this model, thresholds vary for different diseases and at different growth stages, indicating different levels of constraint and sensitivity. Such thresholds may be derived empirically, but it is preferable that they be derived from a knowledge of all constraining factors in models that are at least partially mechanistic. The development of such models and the application of these principles to plants or crops with different yield potentials remains a challenge to pathologists and physiologists.

REFERENCES

ANGUS, J. F., CUNNINGHAM, R. B., MONCUR, M. W., and MACKENZIE, D. H. 1980. Phasic development in field crops. I. Thermal response in the seedling phase. Field Crops Res. 3:365–378.

ARNOLD, S. M., and MONTEITH, J. L. 1974. Plant development and mean temperature in a Teesdale habitat. J. Ecol. 62:711–720.

AUSTIN, R. B., BINGHAM, J., BLACKWELL, R. D., EVANS, L. T., FORD, M. A., MORGAN C. L., and TAYLOR, M. 1980. Genetic improvements in winter wheat yields since 1900 and associated physiological changes. J. Agric. Sci. Camb. 94:675–689.

BATEY, T. 1976. Some effects of nitrogen fertilizer on winter wheat. J. Sci. Food Agric. 27:287–297.

BECKETT, P. H. T., and WEBSTER, R. 1971. Soil variability: a review. Soils Fert. 34: 1–15.

BENNETT, C. M., WEBB, T. M., and WALLACE, A. R. 1980. Influence of soil type on barley yield. N.Z. J. Exp. Agric. 8:111–115.

BOOTE, K. J., JONES, J. W., MISHOE, J. W., and BERGER, R. D. 1983. Coupling pests to crop growth simulators to predict yield reductions. Phytopathology 73:1581–1587.

CHURCH, B. M., and AUSTIN, R. B. 1983. Variability of wheat yields in England and Wales. J. Agric. Sci. Camb. 100:201–204.

COCHRAN, W. G. 1937. A catalogue of uniformity trial data. J. R. Stat. Soc. Ser B 4:233–253.

COCHRAN, W. G. 1977. Sampling Techniques, 3rd ed. John Wiley & Sons, Inc., New York.

COLE, M. J., and GAUNT, R. E. 1987. A disease management response to the introduction of wheat stripe rust to New Zealand. Plant Dis. 71:102–107.

COOPER, J. P. 1975. Control of photosynthetic production in terrestrial systems. *In* Photosynthesis and Productivity in Different Environments (ed. J. P. Cooper), pp. 593–621. Cambridge University Press, Cambridge.

COOPER, P. J. M. 1979. The association between altitude, environment variables, maize growth and yields in Kenya. J. Agric. Sci. Camb. 93:635–649.

COSSENS, G. G. 1983. The effect of altitude on pasture production in the South Island hill country. Proc. Hill and High Country Sem., Lincoln College, New Zealand (ed. B. T. Robertson), pp. 35–40. Centre for Resource Management, Lincoln College, Canterbury, New Zealand.

DAVENPORT, D. C., and HUDSON, J. P. 1967. Local advection over crops and fallow. I. Changes in evaporation rates along a 17 km transect in the Sudan Gezira. Agric. Meteorol. 4:339–352.

DOWDELL, R. J. 1982. Fate of nitrogen applied to agricultural crops with particular reference to denitrification. Philos. Trans. R. Soc. London Ser. B 296:363–373.

ELLIS, R. P., and Kirby, E. J. M. 1980. A comparison of spring barley grown in England and in Scotland. J. Agric. Sci. Camb. 95:111–115.

FAIRFIELD SMITH, H. 1938. An empirical law describing heterogeneity in the yields of agricultural crops. J. Agric Sci. Camb. 28:1–23.

FISHER, R. A. 1924. The influence of rainfall distribution on the yield of wheat at Rothamsted. Philos. Trans. R. Soc. London Ser. B 213:89–142.

FRENCH, B. K., and LEGG, B. J. 1979. Rothamsted irrigation 1964–76. J. Agric. Sci. Camb. 92:15–37.

GAUNT, R. E. 1987. A mechanistic approach to yield loss assessment based on crop physiology. *In* Crop Loss Assessment and Pest Management (ed. P. S. Teng), pp. 150–159. APS Press, St. Paul, Minn.

GOUDRIAAN, J., and VAN LAAR, H. H. 1978. Calculation of daily totals of gross CO_2 assimilation of leaf canopies. Neth. J. Agric. Sci. 26:373–382.

GRIFFITHS, E., and AMIN, S. M. 1977. Effect of *Botrytis fabae* infection and mechanical defoliation on seed yield of faba beans (*Vicia faba*). Ann. Appl. Biol. 86:359–367.

HARPER, J. L. 1977. Population Biology of Plants. Academic Press, Inc. (London) Ltd., London.

HARRIS, J. A., and SCOFIELD, C. S. 1928. Further studies on the permanence of differences in the plots of an experimental field. J. Agric. Res. 36:15–40.

HILLS, F. J., and WORKER, G. F. 1983. Disease thresholds and increases in fall sucrose yield related to powdery mildew of sugar beet in California. Plant Dis. 67:654–656.

HUGHES, K. A., HALL, A. J., GARDER, P. W., KERR, J. P., and WITHERS, N. J. 1984. The prediction of cool season forage oat yield using temperature and solar radiation data. Proc. Agron. Soc. N.Z. 14:65–69.

JAMES, W. C., and TENG, P. S. 1979. The quantification of production constraints associated with plant diseases. *In* Applied Biology, Vol. IV (ed. T. H. Coaker), pp. 201–267. Academic Press, Inc., New York.

JONES, H. G. 1979. Effects of weather on spring barley yields in Britain. J. Natl. Inst. Agric. Bot. 15:24–33.

KARAGEORGIS, D., TONKIN, P. J., and ADAMS, J. A. 1984. Medium and short range variability textural layering in an ochrept developed in an alluvial floodplain. Aust. J. Soil Res. 22:471–474.

KING, J. E. 1976. Relationship between yield loss and severity of yellow rust recorded on a large number of single stems of winter wheat. Plant Pathol. 25:172–177.

KONATO, M. 1984. Climate of the sorghum and millet cultivation zones of semi-arid tropical regions of West Africa. *In* Agrometeorology of Sorghum and Millet in the Semi-arid Tropics. Proceedings of the International Symposium (ed. A. G. R. Virmanish and M. V. K. Sivakumar). ICRISAT, India.

LAMBERT, M. G., and ROBERTS, E. 1976. Aspect differences in an unimproved hill country pasture. N.Z. J. Agric. Res. 19:459–467.

LARCHER, W. 1980. Physiological Plant Ecology, 2nd ed. Springer-Verlag, Berlin.

LIM, L. G., and GAUNT, R. E. 1986. The effect of powdery mildew (*Erysiphe graminis* f. sp. *hordei*) and leaf rust (*Puccinia hordei*) on spring barley in New Zealand. I. Epidemic development, green leaf area and yield. Plant Pathol. 35:44–53.

LOOMIS, R. S. 1983. Productivity of agricultural systems. *In* Encyclopaedia of Plant Physiology, Vol. 12D (ed. O. L. Large, P. S. Nobel, C. B. Osmond, and H. Ziegler), pp. 151–172. Springer-Verlag, Berlin.

LUDLOW, M. M. 1983. External factors affecting photosynthesis and respiration. *In* The Growth and Functioning of Leaves (ed. J. E. Dale and F. L. Milthorpe), pp. 347–380. Cambridge University Press, Cambridge.

LUDLOW, M. M., and WILSON, G. L. 1971. Photosynthesis of tropical pasture plants. Aust. J. Biol. Sci. 24:449–470.

MACKENZIE, D. R., and KING, E. 1980. Development of realistic crop loss models for plant diseases. *In* Crop Loss Assessment, Proc. E. C. Stakman Commemorative Symposium, Miscellaneous Publications, Minnesota Agricultural Experimental Station (ed. P. S. Teng), pp. 85–89, St. Paul, Minn.

MAIN, C. E., and PROCTOR, C. H. 1980. Development of optimal stategies for disease-loss sample surveys. *In* Crop Loss Assessment, Proc. E. C. Stakman Commemorative Symposium, Miscellaneous Publications, Minnesota Agricultural Experimental Station (ed. P. S. Teng), pp. 118–123, St. Paul, Minn.

MARSHALL, J. K. 1967. The effect of shelter on the productivity of grasslands and field crops. Field Crop Abstr. 20:1–14.

MCBRATNEY, A. B. 1985. The role of geostatistics in the design and analysis of field experiments with reference to the effect of soil properties on crop yield. *In* Soil Spatial Variability, Proc. ISSS and SSSA Workshop, pp. 3–8. Soil Science Society of America, Las Vegas, Nev.

MCBRATNEY, A. B., and WEBSTER, R. 1981. Detection of ridge and furrow patterns by spectral analysis of crop yield. Int. Stat. Rev. 49:45–52.

MCCOLLUM, R. E., and MCCABB, S. B. 1954. Chemical properties of some coastal plain soils of North Carolina. Soil Sci. 78:435–443.

MERCER, W. B., and HALL, A. D. 1911. The experimental error of field trials. J. Agric. Sci. Camb. 4:107–132.

MONTEITH, J. L. 1981. Climatic variation and the growth of crops. Q. J. R. Meteorol. Soc. 107:749–774.

MONTEITH, J. L., and SCOTT, R. K. 1982. Weather and yield variation of crops. *In* Food, Nutrition and Climate (ed. K. L. Blaxter and L. Fowden), pp. 127–153. Applied Science Publishers Ltd., Barking, Essex, England.

MYCROFT, H. 1983. Variability of yields in cereal variety × fungicide trials. J. Agric. Sci. Camb. 100:535–538.

PENMAN, H. L. 1962. Woburn irrigation, 1951–59. I. Purpose, design and weather. II. Results for grass. III. Results for rotation crops. J. Agric. Sci. Camb. 58:343–348, 349–364, 365–379.

PENMAN, H. L. 1970. Woburn irrigation, 1960–68. IV. Design and interpretation. V. Results for leys. VI. Results for rotation crops. J. Agric. Sci. Camb. 75:69–73, 75–88, 89–102.

PREW, R. D., CHURCH, B. M., DEWAR, A. M., LACEY, J. PENNY, A., PLUMB, R. T., THORNE, G. N., TODD, A. D., and WILLIAMS, T. D. 1983. Effects of eight factors on the growth and nutrient uptake of winter wheat and on the incidence of pests and diseases. J. Agric. Sci. Camb. 100:363–382.

PREW, R. D., CHURCH, B. M., DEWAR, A. M., LACEY, J., MAGAN, M., PENNY, A., PLUMB, R. T., THORNE, G. N., TODD, A. D., and WILLIAMS, T. D. 1985. Some factors limiting the growth and yield of winter wheat and their variation in two seasons. J. Agric. Sci. Camb. 104:135–162.

RABBINGE, R., and RIJSDIJK, F. H. 1984. Epidemiological and physiological foundation of EPIPRE. *In* Cereal Production (ed. E. J. Gallagher), pp. 215–225. Butterworth & Company (Publishers) Ltd., London.

REDDY, S. J. 1984. Agroclimatic classification of the semi-arid tropics. III. Characteristics of variables relevant to crop production potential. Agric. Meteorol. 30:269–292.

RICHARDSON, M. J., WHITTLE, A. M., and JACKS, M. 1976. Yield-loss relationships in cereals. Plant Pathology 25:21–30.

SAH, D. N., and MACKENZIE, D. R. 1987. Methods of generating different levels of disease epidemics in loss experiments. *In* Crop Loss Assessment and Pest Management (ed. P. S. Teng), pp. 90–96, APS Press, St. Paul, Minn.

SCOTT, R. K., and JAGGARD, K. W. 1978. Theoretical criteria for maximum yield. Proc. 41st Winter Congr. Int. Inst. Sugar Beet Res. (IIRB), pp. 179–198.

SCOTT, R. K., ENGLISH, S. D., WOOD, D. W., and UNSWORTH, M. M. 1973. The yield of sugar beet in relation to weather and length of growing season. J. Agric. Sci. Camb. 81:339–347.

SHANE, W. W., and TENG, P. S. 1987. Generating the database for disease-loss modelling. *In* Crop Loss Assessment and Pest Management (ed. P. S. Teng), pp. 82–89, APS Press, St. Paul, Minn.

SMITH, L. P. 1976. The agricultural climate of England and Wales. MAFF Technical Bulletin HMSO 35.

TENG, P. S. 1987. Quantifying the relationship between disease intensity and yield loss. *In* Crop Loss Assessment and Pest Management (ed. P. S. Teng), pp. 105–113. APS Press, St. Paul, Minn.

TENG, P. S., and SHANE, W. W. 1984. Crop losses due to plant pathogens. CRC Crit. Rev. Plant Sci. 2:21–47.

THOMSON, W. J., and GAUNT, R. E. 1986. The effect of speckled leaf blotch on apical development and yield in winter wheat in New Zealand. Ann. Bot. 58:39–48.

TONKIN, M. H., and SILVEY, V. 1982. Variability of cereal variety trials from fungicide sprayed and unsprayed plots. J. Natl. Inst. Agric. Bot. 16:15–30.

TRANGMAAR, B. B., YOST, R. S., WADE, M. K., UEHARA, G., and SUDJADI, M. 1987. Spatial variation of soil properties and rice yield on recently cleared land. J. Soil Sci. Soc. Am. (in press)

VAN KEULEN, M. 1986. Crop yield and nutrient requirements. *In* Modelling of Agricultural Production: Weather Soils and Crops (ed. M. van Keulen and J. Wolf), pp. 155–181. PUDOC, Wageningen, The Netherlands.

WILLIAMS, G. D. V. 1971. Geographical variations in yield—weather relationships over a large wheat growing region. Agric. Meteorol. 9:265–283.

WILSON, D. R. 1985. The value of water for crop production. N.Z. Agric. Sci. 19:174–179.

WOLEDGE, J., and PARSONS, A. J. 1986. The effect of temperature on the photosynthesis of ryegrass canopies. Ann. Bot. 57:487–497.

YODA, K., KIRA, T., and HOZIMU, K. 1957. Intraspecific competition among higher plants. IX. Further analysis of the competitive interaction between adjacent individuals. J. Inst. Polytech. Osaka City Univ. Ser. 8:161–178.

ZADOKS, J. C. 1984. EPIPRE, a computer-based scheme for pest and disease control in wheat. *In* Cereal Production (ed. E. J. Gallagher), pp. 215–225. Butterworth & Company (Publishers) Ltd., London.

ZADOKS, J. C. 1985. On the conceptual base of crop loss assessment: threshold theory. Annu. Rev. Phytopathol. 23:455–473.

ZADOKS, J. C. 1987. The concept of thresholds: warning, action and damage thresholds. *In* Crop Loss Assessment and Pest Management (ed. P. S. Teng), pp. 168–175, APS Press, St. Paul, Minn.

11
Tree Disease in the Urban Environment
*Spatial Effects and Consequences of **Human** Disturbance*

David N. Appel

INTRODUCTION

Urbanization worldwide has become recognized as an important historical trend during the twentieth century. An inevitable consequence of urbanization is the disturbance of natural ecosystems (Williams et al., 1983). The degree of disturbance varies widely among different locations within a given urban environment (Rowntree, 1984a). In some parts of a community, the ecological diversity of flora and fauna is decreased or completely eliminated due to a dramatic degradation of certain ecosystem components. In contrast, some areas within a community are only minimally influenced by urbanization. The latter areas maintain most of the characteristics of the natural ecosystem in an unaltered state. Between these two extremes are various degrees of complexity and diversity, depending on the level of alteration and social opinion as to what constitutes a suitable environment in which to live. These areas may be delineated as parcels, or discrete small ecosystems, arranged in complex geometric patterns reflecting trends in the urban morphology (Sanders, 1984).

Trees and forests represent one of the most highly valued ecosystem components in the urban environment. Although difficult to quantify in economic

Note: Department of Plant Pathology and Microbiology, Texas A&M University, College Station, TX 77843, US.

terms, the benefits of maintaining an urban forest following development are obvious to many people. In most communities resources are committed, either collectively or individually, to minimize the disturbance of natural forest ecosystems and maintain some level of forestation. These commitments can be considerable when diseases or other natural disasters occur.

An understanding of the dynamics of urban forest ecosystems would prove most useful for studying urban forest diseases (Zadoks and Schein, 1979). Due to the continua of environmental disruption and social influences on urban forest management, any analyses of ecosystem components must contend with numerous unusual complexities (Burns and Manion, 1984). This problem is further compounded when considering a complicated topic such as tree disease epidemiology in an urban context. The management of urban plants is very distinct from traditional agriculture and forest crop production. The direct involvement of numerous agencies, businesses, and landowners makes it difficult to generalize about tree management and the structures of tree populations in towns and cities. The benefits, economics, and environment of urban forestry are also unique. It is therefore undesirable to rely on the extrapolation of results from production agricultural research to the city where the environment is different, the crop mix is distinct, and plant care practices are not even remotely similar (Worf, 1981). Communities will benefit from research and analyses directed toward understanding epidemiology specifically in urban tree populations. Such information assists in developing flexible, efficient systems for the expensive and necessary burden of urban forest management. There are currently very few specific recommendations available to incorporate into tree management decisions for dealing with disease problems in the urban forest.

Observations, and some limited experimental evidence, have provided valuable insights into how urbanization of a natural forest ecosystem and subsequent tree management influences the behavior of tree pathogens. These perceptions will be discussed within the framework of specific diseases considered to be important in urban environments. Obviously, Dutch elm disease has provided the best opportunity to study the affects of the urban environment on the rates and spatial patterns of disease epidemics in urban trees. Additional information can be derived from documentation of diseases such as elm phloem necrosis, dieback of urban maples in the northeast United States, pin oak blight on the eastern shore of Virginia, and the epidemic of oak wilt in central Texas towns and cities. Some of these examples will be compared with the behavior of the same or similar diseases in natural forests to further understand their occurrence in the urban forest.

For the purposes of the following discussion, the urban forest will be defined as all vegetation structures regularly subjected to influences of a distinctly urban nature. This is a process definition of the urban forest which has proven useful for the scientific analyses necessary to develop tenets in such areas as urban forest ecology (Sanders, 1984). Before considering specific disease types or examples, it will be useful to analyze the urban forest from an ecological perspective. In natural ecosystems, forest structure is considered an important influence on the occurrence of epidemics (Schmidt, 1978). Several urban ecosystem typologies have been developed to aid in describing and conceptualizing urban forest structures. These characterizations stratify trends and features of the urban environment into recurring ecological and land use units (Table 1). These units contain predictable urban forest structures (e.g. crown density), and presumably their arrangement will directly influence the spatial and temporal development of an epidemic.

URBAN FOREST ECOLOGY

Some significant regularities may exist in patterns of urban development (Schwirian, 1974) and forest structure for different cities (Sanders, 1984). To predict these patterns, an understanding of processes, both natural and social, responsible for the various configurations is required. Three broad factors are believed to determine urban forest structure. The first, urban morphology, consists of the historical forces in urban development that create spaces available for vegetation. Older cities in the eastern United States, for example, can be expected to have less growing space in inner cities than newer western cities,

TABLE 1 Ecological Land Types, Corresponding Land Use, and Typical Crown Cover Values with the Potential for Epidemic Losses in Urban Environments

Ecological Land Types[a]		Land Use Category[b]		
Urban Land Type	Percent Canopy Cover	Type	Percent Crown Cover	Epidemic Potential
Cliff/Organic Detritus	3	Commercial-industrial	6	Low
Derelict/Weedy Grasslands	3	Vacant	—	Low
Derelict/Savanah	3–50	Vacant	60	High
Mowed/Grassland	30	Institutional-parks	30	Medium
Urban/Savanah	3	Residential	23	Low
Abiotic/Weedy Complex	—	Miscellaneous	16	Low
Urban/Forest Plantation	75	Agricultural/vacant	33	High
Rail-Highway/Grassland	—	Transportation	8	Low
Remnant Ecosystem/Natural Island	50	Parks/vacant	50	High
Remnant Ecosystem/Agricultural Island	—	Agricultural	5	Medium
Lake-Stream/Aquatic Complex	—	—	—	
Dump/Organic Detritus	—	Miscellaneous	16	Low

[a]From Brady et al. (1979), as applied to a portion of southern Ontario.

[b]Values derived from analysis of four eastern U.S. cities (Rowntree, 1984b).

where lower-density urban development has occurred. Natural factors comprise the second group of important influences on urban vegetation. The most important is climatic variation, but such factors as soil types, seed sources, and means of dispersal are also included. In addition to urban morphology and natural factors, human management systems also have an important influence on a city's vegetation. These systems are largely determined by the impact of ownership and land use on the natural ecosystem. If a relationship between land use categories and urban vegetation patterns exists, the spatial arrangement of the land use parcels will certainly exhibit a strong influence on disease epidemics in the urban forest. There is little quantitative evidence available with which to test this proposal, but some modest rules of patterning have been suggested (Sanders, 1984).

Knowledge of land use patterns is particularly important to understanding urban forest epidemiology for a number of reasons. For example, the flow and transmission of pathogen propagules through an urban forest will be influenced by the sizes and distributions of different land use parcels. The opportunity for the development of a tree disease epidemic within individual parcels will also be strongly influenced by land use. Brady et al. (1979) developed a useful ecological typology, in three scales, facilitating the analysis of urban ecosystem dynamics based on land use. Pathogen spread can be considered at each scale to understand the influence of urbanization on an epidemic.

At the largest scale (1:10,000), the urban land system is divided into 12 land types. Each land type is characterized by anthropogenetic and biological elements (Table 1). Although urban morphologies may vary from city to city, the land types and their elements can usually be found and expected to influence epidemics similarly in any urban community. This continuity derives from an association with land use in urban development (Table 1). In some of these land types, the opportunity for pathogen development and significant losses is extremely limited, especially in those where trees are nonexistent or nearly so. These would include such land types as the Lake-Stream/Aquatic Complex (LS/AC) or the Dump/Organic Detritus (D/OD) (Table 1). There are few trees found in these land types, and the consequences of diseases are minimal. The same will sometimes be true of the Cliff/Organic Detritus (C1/OD) land type, where high vertical buildings, narrow streets, and pervasive pavement eliminate growing space and prevent canopy cover from exceeding more than 3%. In some cities, however, special provisions are made for large numbers of trees, often exotics planted in containers or holes in the pavement. When present in large numbers, the possibility of localized epidemics and potential as sources of inoculum for adjacent land types should not be ignored. It should also be recognized that the vertical distribution of buildings, dominance of pavement, and pollution load found in the C1/OD land type will influence the climates of adjacent land types and in turn may influence the prevalence of various diseases.

In contrast to those land types mentioned above, certain areas in the urban environment are dominated by trees and will have a strong influence on tree disease epidemics. Two examples, with very different forest structures, are the Urban/Savanah (U/S) and Remnant Ecosystem/Natural Island (RE/NI) land types. Each of these has the potential for a high proportion of canopy cover, but species composition and tree distributions can be very different. These differences are important, as demonstrated by the rapid spread of Dutch elm disease in communities where American elm was planted as a monoculture in evenly spaced plantings. These land types will be considered further in the discussion of specific diseases.

An important issue in urban plant ecology, concerning species diversity, has a direct application to principles in urban tree epidemiology. A long-held ecological tenet associates high species diversity with ecosystem stability (Guntenspergen and Stearns, 1983; Richards, 1983). This apparent relationship is being used to recommend increasing diversity during planting. Municipal forestry departments have adopted a policy of planting no more than 10 to 15% of the total population to a single species (Grey and Deneke, 1986). High species diversity is believed to contribute stability to a street tree population by preventing catastrophic losses of a single species from such hazards as insects or diseases. This logic is strongly reinforced by the North American Dutch elm disease epidemic. The widespread losses incurred certainly proved the vulnerability of a street tree monoculture to a virulent pathogen. Similar vulnerabilities have been repeatedly proven in traditional agroecosystems, and in this respect urban forests are apparently no different.

There are, however, problems in assuming that long-term stability in a plant population can be gained simply by increasing species diversity (Zadoks and Schein, 1979). Some problems will arise from the unique differences between urban forests and natural forest ecosystems. The complexities in structure and form, as well as the imposition of intensive vegetative management systems, alter forest ecosystems. It is reasonable to assume that ecological concepts derived from natural forests must also be changed. Management objectives for the urban forest are largely different from those of most agroecosystems, making comparisons to even those artificial systems difficult. Richards (1983) argues that arbitrarily increasing the species diversity of the urban forest may actually decrease stability of the population from an urban forestry standpoint.

SPECIFIC DISEASE TYPES

Vascular Wilts

The wilts are caused by organisms, mostly fungi, that disrupt xylem conduction in the translocation stream, usually resulting in rapid wilting and death of the infected tree (Blanchard and Tattar, 1981). There are numerous examples of particularly damaging vascular wilt diseases on urban tree species. These include Dutch elm disease, caused by *Ceratocystis ulmi;* oak wilt, caused by *C. fagacearum;* Verticillium wilt of hardwoods (*Verticillium dahliae*); and mimosa wilt of *Albizzia vulibrissin,* caused by *Fusarium oxysporum.* These diseases have attained notoriety in urban forestry for different reasons. In certain cases, planting practices have increased the vulnerability of the urban forests to widespread losses. Also, some of these vascular wilt pathogens are uniquely suited for rapid spread or long persistence in urban soils.

The Dutch elm disease pathogen has two means of transmission. One is by means of contaminated bark beetles that emerge from dead and dying trees to carry the fungus to new trees during feeding. The other means of transmission is through functional root grafts formed between adjacent trees. Following multiple introductions of *C. ulmi* into North America, the progression of elm mortality was recorded for numerous northeastern and midwestern communites in the United States (Miller et al., 1969; Neely, 1967; Schmid, 1975; Schread et al., 1946). The occurrence of Dutch elm disease has also been described for various European communities (Gibbs, 1978). Unfortunately, most of these reports concern temporal progress of losses, or control efficacy on a community basis, with little analysis of the spatial distribution of the disease. One notable exception is a detailed description of the occurrence of Dutch elm disease in the cities of Champaign and Urbana, Illinois (Carter and Carter, 1974).

The first elm affected by Dutch elm disease in these adjacent Illinois cities was found in southwest Urbana in 1951. In 1952, 11 elms were killed, and 176 were affected in 1953. During this three-year period, only one tree was within root-grafting distance of another wilting elm, implicating elm bark beetles as responsible for spread. From the single infection in 1951, the fungus spread rapidly throughout much of the Champaign–Urbana area. By 1972, the final year of the survey, 11,062 trees were killed by Dutch elm disease. There was no mention of any trends in fungal spread with regard to the structure of the elm population on a city-wide basis. The early spread of Dutch elm disease was believed to be influenced by the presence of another disease, elm phloem necrosis, probably because of the large numbers of bark beetles occurring in areas where trees died from the mycoplasma infections.

When analyzed on a smaller local scale along individual streets, less than half of the trees were suspected to be infected by root grafting between adjacent trees. For example, a detailed description is given (Carter and Carter, 1974) of disease progression on both sides of a street along one city block in Urbana where 28 trees were killed during 1955–1959. Following the initial infections of 3 trees in the first year on both sides of the street, 15 of the trees became diseased at distances too far to be attributed to root grafting.

The description of Dutch elm disease in Champaign and Urbana is typical of the accounts of that disease in other cities. Emphasis is placed on residential neighborhoods where planted "parkway" elms comprised 50% or more of the tree populations, usually planted ajacent to one another in rows. Under these conditions, *C. ulmi* has consistently demonstrated a propensity for rapid spread. These conditions are largely those found in the U/S urban ecological land type (Table 1). Residential zones are usually considered to comprise the majority of trees in a community, often to the extent of excluding other land types with lower tree densities in urban forestry analyses (Mohai et al., 1978).

Tree density has been identified as the most important single factor affecting Dutch elm disease progress in natural stands. Dense aggregations of elms in natural forests increase the intensity and spread of infection (Pomerleau, 1961). In towns where elm trees are more scattered, a slower rate of spread has been observed than in those towns with high concentrations of elms. Other factors have been identified that would influence the occurrence of Dutch elm disease in different land types. Pruning cycles maintained by municipal tree maintenance personnel were also found to influence disease distribution in Detroit, Michigan (Hart et al., 1967). Under these circumstances, the periodic pruning cycle would certainly have a direct bearing on the spatial distribution of the disease, and it is assumed that pruning cycles maintained by a municipality may be associated with land types and land use. Soil type, soil moisture-holding capacity, and available soil moisture are all edaphic factors known to influence the susceptibility of American elm (*Ulmus americana*) to *C. ulmi* (Kais et al., 1962). All of these factors will vary in a given urban environment according to land use and degree of disturbance.

The oak wilt epidemic in central Texas towns and cities provides another good example for analyzing urban tree epidemiology. Although some important differences exist between semi-evergreen live oaks (*Quercus fusiformis*) affected by oak wilt and American elm, tree density and other aspects of urban forest structure have a major influence on the distribution of the disease.

Ceratocystis fagacearum, the oak wilt pathogen, is transmitted by root connections between adjacent trees in a manner similar to *C. ulmi*. The oak wilt pathogen also has an insect vector, sap-feeding nitidulid beetles. Oak wilt is particularly serious in the U/FP ecosystems of central Texas, where urban development was designed to preserve large, homogeneous stands of live oak. An important characteristic of live oak is vegetative reproduction through root sprouting. This habit appears to provide an efficient connection for *C. fagacearum* colonization among native trees, resulting in rapidly expanding infection foci. The fungus spreads between adjacent trees in urban environments at rates comparable to those observed under natural conditions. In many cases, streets, building foundations, and underground utilities have proven ineffective in impeding transmission of the fungus.

Unlike the Dutch elm disease fungus, the oak wilt pathogen is not efficiently vectored by insects from a majority of diseased trees. The oak wilt fungus forms a discrete fungal mat beneath the bark of only diseased red oaks (genus *Quercus* sub gen. *Erythrobalanus*). No fungal mats have been found forming on white oaks (sub gen. *Leucobalanus*) or on live oaks. In central Texas, fungal mats form occasionally on Spanish oaks (*Q. texana*) when environmental conditions are favorable and the diseased tree is in the right stage of colonization (Appel et al., 1987). In the urban environment, mat-bearing Spanish oaks are most often found in empty lots and parks, characteristic of the RE/NI ecological land types. Spanish oaks usually comprise only a small proportion of the total oak population in natural forests, and are present in even smaller numbers following urbanization.

Although vascular wilt diseases other than Dutch elm disease and oak wilt are less spectacular, they exhibit characteristics of great importance to their spatial distribution in the urban environment. *Verticillium dahliae*, for example, causes a vascular wilt of elms and maples of local importance in communities in the northeastern and midwestern United States. The Verticillium wilt pathogen has a very wide host range on numerous tree species and agronomic crops (Manion, 1981). Unlike the oak wilt and Dutch elm disease pathogens, *V. dahliae* is able to persist for long periods of time in the soil and enter trees through wounds in roots. Urban development on former agricultural land where high populations of *V. dahliae* have developed may encourage disease problems on newly planted trees. Many other soil-inhabiting pathogens are also able to make the transition from agricultural crops to transplanted urban trees, emphasizing the importance of site history in urban tree epidemiology.

Declines

Tree declines are not unique to the urban environment, but they comprise some of the most serious problems in urban forest management (Gilbertson and Bradshaw, 1985; Howe, 1974; Rice, 1976). Declines are generally considered to be the result of numerous interacting factors, such as too little or too much soil moisture, nutrient deficiencies, soil compaction, lack of growing space, mechanical damage, improper transplanting, or pests and pathogens that normally do trees no harm (Manion, 1981; Sinclair, 1966). Although the factors and species involved may vary from one urban community to another, the symptoms of different declines are fairly consistent.

A concept of declines has evolved to explain how the various different factors interact to result in tree mortality (Manion, 1981; Sinclair, 1966). The concept is derived by categorizing the determinants of declines into three groups. The first group is comprised of abiotic, *predisposing* factors, such as long-term drought or site deterioration. They impose subtle physiological stresses in urban environments, where trees are often subjected to conditions not encountered in natural forest ecosystems. The next group, *inciting* factors, may be abiotic or biotic. Examples include construction damage, acute air pollution damage, or periodic defoliation by insects. Tree health degenerates rapidly following exposure to inciting factors. Finally, *contributing* factors such as borers, root rotters, or canker fungi attack or colonize the weakened tree, resulting in death over a period of years.

Maple decline has been frequently studied and described for urban communities in the northeastern and midwestern United States (Burns and Manion, 1984; Kielbaso and Ottman, 1976; Dyer and Madar, 1986). For example, the spatial distribution of declining maples was the subject of a study during 1975–1977 in two residential neighborhoods of Syracuse, New York (Burns and Manion, 1984). Symptomatic maples exhibiting large dead limbs, small dead limbs, scorch, or chlorosis were found to be aggregated, sometimes along the street of an entire city block. Trees with chlorosis were aggregated in concentrated circular patterns, while the other three symptoms were found in linear patterns, sometimes extending the length of several blocks. These patterns varied when compared among different species of maples.

The spatial patterns in symptoms of maple decline in Syracuse were believed to be influenced by street-related and localized site-related factors. No attempt was made to identify specific factors, but biotic pathogens were for the most part believed to be necessarily linked to environmental factors. The same conclusion was reached in an analysis of pin oak decline and blight in the *Quercus palustris*-dominated urban forests of eastern Virginia (Appel and Stipes, 1986). *Endothia gyrosa*, the pin oak blight pathogen, produces cankers on trees and is associated with pruning wounds (Hunter and Stipes, 1978). The fungus, however, does not appear to infect and colonize wounds unless potential hosts are predisposed by some environmental factor such as water stress (Appel and Stipes, 1984). A survey and inventory of 580 pin oaks in a residential, industrial community in eastern Virginia was conducted to determine whether concrete overlay or the presence of underground utility systems may also influence disease incidence (Appel, 1980). The health of trees did not seem to deteriorate with an increasing number of underground utilities. Also, a decrease in the diameters of trees near three utility systems, as opposed to none, was not sufficiently large to be statistically significant (Table 2). A propensity for heavy cankering was found to correlate with slower rates of growth among similarly aged individuals. The average reduced growth rates for trees with one to three underground utilities may reflect the inevitable maintenance and repair activities within the root zones of the trees. However, there is no direct evidence to explain the incidence of pin oak blight with the occurrence of specific environmental factors. As with numerous other tree declines, severely declining and blighted pin oaks were commonly found adjacent to trees exhibiting no symptoms of decline.

The spatial distributions of declines will probably be distinct from those observed for virulent pathogens in urban environments. Declines will often occur and be noticed where trees, allowed to remain following development, are exposed to severe stress. In this sense, the distribution of a decline is actually a symptom of where natural vegetation will have difficulty surviving. If sufficiently severe, environmental stresses cause tree predisposition toward in-

TABLE 2 Average Diameters at Breast Height (dbh) and Health Ratings for Groups of Trees with Root Systems Intercepted by Underground Utilities in an Eastern Virginia Residential Community

Numbers of Utilities[a]	Numbers of Trees	Average Health Rating[b]	Average dbh (cm)
0	88	8	56.3
1	72	8	53.4
2	65	8	53.2
3	32	8	53.3

[a]The utilities included electrical distribution, water distribution, and/or sewage removal systems.

[b]Health was assessed subjectively by leaf color, proportion of crown dieback, and numbers of *Endothia* cankers on a scale of 1 through 10 (1, dead; 10, healthy).

creased susceptibility to nonaggressive, facultative saprophytes (Appel and Stipes, 1984). This involvement of a "secondary" pathogen as a contributing factor adds to the complexity of influences potentially involved in a decline. Inoculum availability, climatic factors, and potential vectors must also be considered in evaluating the epidemiology of many declines.

Miscellaneous Diseases

Urbanization will have a significant influence on the occurrence of many tree foliar pathogens. This influence arises from the impact of structures and artificial surfaces on climatic variables. Also, the resultant tree structure following urban development will have a great influence on the incidence and severity of foliar diseases. Foliage diseases are particularly affected by available moisture, temperature, wind, and light (Kessler, 1981). All of these factors will be modified by paved streets, parking lots, and placements of buildings. In some cases, the lack of foliar pathogens in urban tree plantings may be attributed to reduced periods of leaf wetness (Kessler, 1981). The spatial relations of trees to one another and to topographical features of the landscape will also be important factors in the development of foliar diseases. Features contributing to high humidity and shading will be conducive to foliar pathogens.

An additional consideration of urban forest diseases is wounding. The degree of wounding sustained by a population of trees can have a direct effect on the occurrence of certain diseases. Many canker, heartrot, and root rot organisms require wounds as infection courts and can be expected to increase in incidence where wounding is prevalent. Wounds can be created by vandalism, negligent driving, construction, or during routine tree maintenance. The forests in each of the land types in Table 1 will at sometime be subjected to wounding, probably greater than those levels sustained in the natural forest

ecosystem. The timing and degree of wounding will then influence the levels of a particular disease.

CONCLUSIONS

A virulent pathogen causing an epidemic in the urban forest can have a dramatic, detrimental impact on the value of urban property. For example, this is happening in the numerous central Texas towns and communities sustaining oak wilt. The available approaches for direct oak wilt control are either inappropriate or ineffective when used in the urban environment. A better understanding of the influence of forest structure on spread of *C. fagacearum* in live oaks may lead to some practical cultural practices for disease control, such as selective thinning of dense stands. These sorts of practices would be more useful if they were inexpensive and could be incorporated easily into existing methods for planning and development of urban property. It must be recognized that the structure of the urban forest will often be given a low priority in urban planning, especially if there is no certainty of achieving anticipated benefits from the costs of implementing recommendations.

In addition to helping protect future urban forests, there is merit in understanding the spatial and temporal characteristics of urban tree diseases to protect our existing urban forests. A better understanding of species composition, for example, could have a beneficial influence on available management practices and may also improve the appearance of our urban environments at little cost. Again, these benefits can only be gained from a better understanding of urban tree epidemiology.

Although certain diseases have been documented in detail, there is very little quantitative information available for drawing conclusions about urban forest structure and influences on tree diseases. In this regard the concepts of studies in urban forest ecology may provide a useful framework for advances in urban forest epidemiology.

REFERENCES

APPEL, D. N. 1980. The influence of selected urban site factors, host nutrition and water stress on the decline and blight of pin oak incited by *Endothia gyrosa*. Ph.D. dissertation, Virginia Polytechnic Institute and State University.

APPEL, D. N., and STIPES, R. J. 1984. Canker expansion on water stressed pin oaks colonized by *Endothia gyrosa*. Plant Dis. 68:851–853.

APPEL, D. N., and STIPES, R. J. 1986. A description of declining and blighted pin oaks in eastern Virginia. J. Arboric. 12:155–158.

APPEL, D. N., PETERS, R., and LEWIS, R. 1987. Tree susceptibility, inoculum availability, and potential vectors in a Texas oak wilt center. J. Arboric. 13:169–173.

BLANCHARD, T. A., and TATTAR, R. O. 1981. Field and Laboratory Guide to Tree Pathology. Academic Press, Inc., New York.

BRADY, R. F., TOBIA, T., EAGLES, P. F. J., OHRNER, R., MICAK, J., VEALE, B., and DORNEY, R. S. 1979. A typology for the urban ecosystem and its relationship to larger biogeographical landscape units. Urban Ecol. 4:11–28.

BURNS, B. S., and MANION, P. D. 1984. Spatial distribution of declining urban maples. Urban Ecol. 8:127–137.

CARTER, J. C., and CARTER, L. R. 1974. An urban epiphytotic of phloem necrosis and Dutch elm disease, 1944–1972. Ill. Nat. Hist. Surv. Bull. 31:113–143.

DYER, S. M., and MADAR, D. L. 1986. Declined urban sugar maples: growth patterns, nutritional status, and site factors. J. Arboric. 12:6–13.

GIBBS, J. N. 1978. Intercontinental epidemiology of Dutch elm disease. Annu. Rev. Phytopathol. 16:281–307.

GILBERTSON, P., and BRADSHAW, A. D. 1985. Tree survival in cities: the extent and nature of the problem. Arboric. J. 9:131–142.

GREY, G. W., and DENEKE, F. J. 1986. Urban Forestry, 2nd ed. John Wiley & Sons, Inc., New York.

GUNTENSPERGEN, G., and STEARNS, F. 1983. Comments on N. A. Richards' diversity and stability in a street tree population. Urban Ecol. 7:173–176.

HART, J. H., WALLNER, W. E., CARIS, M. R., and DENNIS, G. K. 1967. Increase in Dutch elm disease associated with summer trimming. Plant Dis. Rep. 51:476–479.

HOWE, V. K. 1974. Site changes and root damage: some problems with oaks. Morton Arbor. Q. 10:49–53.

HUNTER, P. P., and STIPES, R. J. 1978. The effect of month of inoculation with *Endothia gyrosa* on development of pruned branch cankers of pin oak (*Quercus palustris*). Plant Dis. Rep. 62:940–944.

KAIS, A. G., SMALLEY, E. B., and RIKER, A. J. 1962. Environment and development of Dutch elm disease. Phytopathology 52:1191–1196.

KESSLER, K. J. 1981. Considerations of microclimate-leaf disease relations in arboriculture. J. Arboric. 7:169–173.

KIELBASO, J. J., and OTTMAN, K. 1976. Manganese deficiency contributory to maple decline? J. Arboric. 2:27–32.

MANION, P. D. 1981. Tree Disease Concepts. Prentice-Hall, Inc., Englewood Cliffs, N.J.

MILLER, H. C., SILVERBORG, S. B., and CAMPANA, R. J. 1969. Dutch elm disease: relation of spread and intensification to control by sanitation in Syracuse, New York. Plant Dis. Rep. 53:551–555.

MOHAI, P., SMITH, L., VALENTINE, F., STITELER, W., ELIAS, T., and WESTFALL, R. 1978. Structure of urban street tree populations and sampling designs for estimating their parameters. *In* METRIA: 1, Proc. Metro. Tree Impr. Alliance (ed. F. S. Santamour, jr.), pp. 28–43. U.S. For. Serv. State and Private Forestry, Northeastern Area.

NEELY, D. 1967. Dutch elm disease in Illinois cities. Plant Dis. Rep. 51:511–514.

POMERLEAU, R. 1961. History of the Dutch elm disease in the province of Quebec, Canada. For. Chron. 37:356–67.

RICE, P. F. 1976. Urban tree pathology of human settlements—a Canadian experience. *In* Trees and Forests for Human Settlements (ed. J. W. Andersen), pp. 321–33. Proc. Pl. 05-00 Symp. Arboriculture Urban For. Centre for Urban Forestry Studies, University of Toronto, Toronto, Canada.

RICHARDS, N. A. 1983. Diversity and stability in a street tree population. Urban Ecol. 7:159–171.

ROWNTREE, R. A. 1984a. Ecology of the urban forest—introduction to Part I. Urban Ecol. 8:1–11.

ROWNTREE, R. A. 1984b. Forest canopy cover and land use in four eastern United States cities. Urban Ecol. 8:55–67.

SANDERS, R. A. 1984. Some determinants of urban forest structure. Urban Ecol. 8:13–27.

SCHMID, J. A. 1975. Urban vegetation: a review and Chicago case study. Dep. Geogr. Univ. Chicago Res. Pap. 161. 266 pp.

SCHMIDT, R. A. 1978. Diseases in forest ecosystems: the importance of functional diversity. *In* Plant Diseases: An Advanced Treatise, Vol 2 (ed. G. Horsfall and E. B. Cowling), pp. 287–315. Academic Press, Inc., New York.

SCHREAD, J. C., WALLACE, P. R., and ZENTMYER, G. A. 1970. Dutch elm disease in southwestern Connecticut. Conn. Agric. Exp. Stn. Bull. (New Haven) 501:33–34.

SCHWIRIAN, K. P. (ed.) 1974. Comparative Urban Structure. D.C. Heath and Company, Lexington, Mass.

SINCLAIR, W. A. 1966. Decline of hardwoods: possible causes. Proc. Int. Shade Tree Conf. 42:17–32.

WILLIAMS, J. F., BRUNN, S. D., and DADEN, J. T. 1983. World urban development. *In* Cities of the World (ed. S. D. Brunn and J. F. Williams), pp. 3–41. Harper & Row Publishers, Inc., New York.

WORF, G. L. 1981. Does urban plant pathology have a future? Plant Dis. 65:470–474.

ZADOKS, J. C., and SCHEIN, R. D. 1979. Epidemiology and Plant Disease Management. Oxford University Press, Inc., New York.

Index

C

D

E

F

G

H